Work and Technological Change

The Clarendon Lectures in Management Studies are jointly organized by Oxford University Press and the Saïd Business School. Every year a leading international academic is invited to give a series of lectures on a topic related to management education and research, broadly defined. The lectures form the basis of a book subsequently published by Oxford University Press.

Clarendon Lectures in Management Studies:

The Modern Firm
Organizational Design for Performance and Growth
John Roberts

Managing Intellectual Capital
Organizational, Strategic, and Policy Dimensions
David Teece

The Political Determinants of Corporate Governance
Political Context, Corporate Impact
Mark Roe

The Internet Galaxy
Reflections on the Internet, Business, and Society
Manuel Castells

Brokerage and Closure
An Introduction to Social Capital
Ronald S. Burt

Reassembling the Social
An Introduction to Actor-Network-Theory
Bruno Latour

Science, Innovation, and Economic Growth
Walter W. Powell

The Logic of Position, the Measure of Leadership
Position and Information in the Market
Joel Podolny

Global Companies in the 20th Century
Leslie Hannah

Gatekeepers
The Role of the Professions in Corporate Governance
John C. Coffee

Material Markets
How Economic Agents Are Constructed
Donald MacKenzie

Corporations in Evolving Diversity
Cognition, Governance, and Institutions
Masahiko Aoki

Staying Power
Six Enduring Principles for Managing Strategy and Innovation in an Uncertain World
Michael A. Cusumano

The Entrepreneurial Firm: Strategy and Organization in New Markets
Kathleen M. Eisenhardt

Doing New Things in Old Organizations
The (Business) Challenge of Climate Change
Rebecca M. Henderson

Maverick Markets
The Virtual Societies of Financial Markets
Karin Knorr Cetina

Disruptive Innovation and Growth
Clayton Christianson

The Architecture of Collapse
The Global System in the 21st Century
Mauro F. Guillén

The 99 Percent Economy
How Democratic Socialism Can Overcome the Crises of Capitalism
Paul S. Adler

Work and Technological Change
Stephen R. Barley

Selves at Work
Herminia Ibarra

Work and Technological Change

Stephen R. Barley

OXFORD
UNIVERSITY PRESS

OXFORD
UNIVERSITY PRESS

Great Clarendon Street, Oxford, OX2 6DP,
United Kingdom

Oxford University Press is a department of the University of Oxford.
It furthers the University's objective of excellence in research, scholarship,
and education by publishing worldwide. Oxford is a registered trade mark of
Oxford University Press in the UK and in certain other countries

Published in the United States of America by Oxford University Press
198 Madison Avenue, New York, NY 10016, United States of America

British Library Cataloguing in Publication Data
Data available

Library of Congress Control Number: 2020944348

ISBN 978-0-19-879520-9

DOI: 10.1093/oso/9780198795209.001.0001

Preface

I was and remain honored to have been selected to deliver the Clarendon Lectures at Oxford University in 2016. Looking over the list of those who gave these lectures before me was humbling, and I still wonder if someone did not make a mistake. Regardless, I am grateful to the Oxford University Press and to the Saïd School of Business for deeming me worthy of the honor.

For my lectures, I chose to write three essays (Chapters 1, 2, and 4) that summarize what I think I have learned over nearly forty years as student of technology and work and of the technical workforce. The fourth essay (Chapter 3) was written after delivering the lectures. I intend the essays and this book to bring closure to my career to date. I am not sure whether I will continue to study technology and work in the future, but if I do, I suspect that whatever I write will stand on the thoughts and conclusions found among these pages. It is hard to teach old dogs new tricks, and with each passing day I become an older dog.

The first essay—"What Is a Technological Revolution?"—presents the fruits of my struggle to come to terms with the broad history of technological change and with where current developments fit into the scheme by which I have come to make sense of the First, the Second, and what I will call the Control Revolution. The last involves computers, microprocessors, robotics, sensors, machine learning, and algorithms, which are again making technology and work a topic of current concern. The level of analysis here is far more sweeping than the level of analysis for which my work is known, since I am primarily an ethnographer and not a historian or a theorist. The essay is rooted in lectures that I developed for my undergraduate students at Stanford with the hope that they might be better able to put the era in which they live into perspective by linking the past to the present. If nothing else, the essay gave me the opportunity to retell Pertti Pelto's (1973) story of *The Snowmobile*

Revolution: Technology and Social Change in the Artic. Pelto's book influenced me strongly when I was a graduate student, and to this day I count it as one of the best ethnographies ever written on the social consequences of technological change. The essay also allows me to acquaint or reacquaint scholars with the work of William Faunce and to initiate a discussion of intelligent technologies and the so-called Fourth Industrial Revolution which the third essay pursues in greater depth.

The second essay—"How Do Technologies Change Organizations?"—brings me back to my comfort zone. The essay recounts what I think I have learned about how technologies change work and about when those changes will and will not result in organizational change. Specifically, I propose a role-based theory of technologically occasioned organizational change that I began to develop in my earliest work on radiology and that I have continued to elaborate since then. Unsurprisingly, for those who have read some of my work, my view is rooted in Industrial Sociology and, in particular, the studies of work done by Chicago School Sociologists of the 1940s and 1950s, long my heroes. Among them I count Everett C. Hughes, Howard Becker, and Anselm Strauss, all of whom I was fortunate enough to meet in my lifetime.

I wrote the third essay—"How Should We Study Intelligent Technologies' Implications for Work and Employment?"—with my colleague at UCSB, Matt Beane, who knows more about intelligent technologies and how they operate than I do. The essay lays out our notions of how researchers can better investigate the ramifications of intelligent technologies to answer the question "How are intelligent technologies likely to change the nature of work and employment?" This question is currently a topic of great discussion, debate, consternation, and fear as we head deeper into the twenty-first century. The essay draws on ideas discussed in the first and second essays. It identifies and discusses two reasons why we see the current state of research and speculation on intelligent technologies in the workplace as inadequate.[1] The essay also provides pointers to the

---

[1] See Bailey and Barley (forthcoming) for additional thoughts on what it will take to strengthen our current understanding of how intelligent technologies may affect work and employment.

kinds of research that would be useful for assessing more clearly what intelligent technologies, especially artificial intelligence, may do to work and employment. Unless we produce not only more but better empirical studies, we are likely to stumble our way into a future that the majority of us may or may not want. I do not believe that sociotechnical trajectories are foreordained, either for reasons of progress or a technology's inevitable unfolding. We always have a choice if we (or, more accurately, powerful people) can muster the data and the will to make a informed choices for the benefit of the majority.

I wrote the fourth essay with my friend and long-time colleague Diane Bailey. She and I have spent most of our careers studying professional and, particularly, technical workers. Technical work is difficult to study. Over the years Diane and I developed methods and approaches that helped us study technical work even though we often knew relatively little about what our informants did or how they did it at the time we set off into the field. "Managing the Fears of Studying Technical Work" lays out the uncertainties and trepidations that led us as fieldworkers to invent methods for studying technical work that allowed us to assuage our fears. Our hope is that this essay will be of some use to young ethnographers who also decide to study technical work and technical occupations.

I have chosen to write a book of essays rather than a book that moves from chapter to chapter developing a single argument or thesis for several reasons. First, I have preferred the genre of the essay from the time I was an English major at William and Mary. In an essay you can lay out a position without having to drive home your point in excruciating and boring detail. I hope we have succeeded on this score, but only readers can tell. Second, if my colleagues and I say anything that other scholars find useful, they can more easily assign or recommend essays than they can books to their students in today's environment, where students, at least, resist reading books. I suspect this is one of the upshots of the Internet, social media, and television, which present information in easily digestible, but perhaps not fully nourishing bites. Third, once I know what I want to say, I do not usually have the patience or stamina to drag it out.

A book like this, intended to culminate a body of scholarship conducted over many years in a variety of settings, owes much to those with whom I have collaborated. In addition to Diane and Matt, I am grateful for the colleagueship, insights, and skills of those with whom I have written: Beth Bechky, Asaf Darr, James Evans, John Freeman, Stine Grodal, Ece Gursoy, Ralph Hybels, Gideon Kunda, Paul Leonardi, Debra Meyerson, Bonnie Nelsen, Siobhan O'Mahony, Wanda Orlikowski, Julian Orr, Hatim Rahman, Mario Scarselletta, Pam Tolbert, Peter Whalley, and Stacia Zabusky. All worked with me on one or more of the projects that helped me learn what I think I know about technological change and technical work. Each greatly influenced my thinking. I am also grateful to colleagues who have not been my coauthors, but who provided intellectual and moral support and who helped me in one way or another sharpen my ideas and avoid the many morasses into which all scholars risk falling: Mark Granovetter, Pam Hinds, Kyle Lewis, Peter Manning, Woody Powell, Rene Rottner, Bob Stern, and Bob Sutton, All idiotic ideas are my own. Thanks also to Chris Wardmann from Hybrid Digital for helping Matt and me flesh out the material and digital stacks for Affdex in Chapter 3.

I also owe a considerable debt to John Van Maanen who taught me how to do ethnography and to Tom Allen who turned my interest toward studying technology and technical workers when I was a doctoral student. They, along with Lotte Bailyn and Ed Schein, made my doctoral studies an adventure in the freedom of thought.

Last, but not least, I owe over four decades of gratitude to my wife Debbi. She, more than anyone, has had to put up with me over the years. I was very fortunate to have found her when we were undergraduates and even more fortunate that she has stuck by me through thick and thin.

Stephen R. Barley
Christian A. Felipe Professor of Technology
Management
Technology Management Program
College of Engineering
University of California, Santa Barbara

January, 2020

Contents

List of Figures   xi

1. **What is a Technological Revolution?**   1

2. **How Do Technologies Change Organizations?**   25
   Roles, Role Relationships, and Encounters   28
   Computerized Imaging   36
   Selling Cars through the Internet   52
   Discussion   62

3. **How Should We Study Intelligent Technologies'**
   Implications for Work and Employment?   69
   Current Thinking on Intelligent Technology, Work,
     and Employment   71
   A Technologically Embedded and Role Systems View
     of Railroads   80
   A Technologically Embedded and Role Systems View
     of Intelligent Technologies   86
   Implications for Future Research on Intelligent
     Technologies and Work   103
   Epilogue   112

4. **Managing the Fears of Studying Technical Work**   116
   Fear #1: Looking Stupid   122
   Fear #2: What Did They Just Say?   124
   Fear #3: No lo comprendo   127
   Fear #4: How Will We Capture the Complexity?   129
   Fear #5: How Will I Finish the Fieldwork Before I Die?   134
   Fear #6: How Will I Make Sense of All These Data?   140
   Conclusion   142

References   145
Index   155

List of Figures

1.1. Infrastructural Technologies Occasion Change in Work
That Ramify Across the Institutional Sectors of Society   9

1.2. Faunce's Theory of Technological Revolutions   17

2.1. How Technologies Change Organizations   27

2.2. The Dramaturgical Elements of an Encounter   33

2.3. Role Structure of a Radiology Department According to
Job Descriptions   38

2.4. Script for Doing a Routine X-Ray   41

2.5. Script for Doing a Sonogram   45

2.6. Networks of Work Relations at Suburban and Urban Hospitals 51

2.7. The Floor Script   54

2.8. The Internet Script   59

3.1. Framework for Embedded Analysis of Technology   82

3.2. Technical Base for Passenger Rail   83

3.3. Material Technical Base for Affdex   89

3.4. Digital Technical Base for Affdex   93

3.5. Role Systems Analysis of an Intelligent Technology   102

4.1. Barley's Matrix Design for Studying Technicians' Occupations 137
After S.R. Barley (1996) "Technicians in the Workplace: Ethnographic
Evidence for Bringing Work into Organization Studies,"
Administrative Science Quarterly, 41: 404–41.

1
What is a Technological Revolution?

Recently a growing number of commentators, scholars, and technological evangelists (located primarily, but not exclusively, in the Silicon Valley) have begun to promote the idea that we have entered an era that will become known as the Fourth Industrial Revolution. This catchy idea is beginning to spread. For example, Klaus Schwab, founder and executive chairman of the World Economic Forum, writes:

> We stand on the brink of a technological revolution that will fundamentally alter the way we live, work, and relate to one another. In its scale, scope, and complexity, the transformation will be unlike anything humankind has experienced before. We do not yet know just how it will unfold, but one thing is clear: the response to it must be integrated and comprehensive... The First Industrial Revolution used water and steam power to mechanize production. The Second used electric power to create mass production. The Third used electronics and information technology to automate production. Now a Fourth Industrial Revolution is building on the Third... It is characterized by a fusion of technologies that is blurring the lines between the physical, digital, and biological spheres. (Schwab, 2016)

In general, those who speak about the Fourth Industrial Revolution point to intelligent technologies which, among other things, include advances in artificial intelligence and machine learning, self-driving cars, drones, big data, the Internet of things, the increasing ubiquity of smartphones with their seemingly endless array of apps, and

Work and Technological Change. Stephen R. Barley, Oxford University Press (2020). © Stephen R. Barley.
DOI: 10.1093/oso/9780198795209.003.0001

ultimately the fusion of the digital and the biological to create cyborgs. All of these technologies are likely to be incredibly important, but unlike advocates of the Fourth Industrial Revolution, I see these developments as rooted in and building upon what I will call the control revolution. My intention, however, is not to debate the pros and cons of how many technological revolutions Western society has experienced or will experience. Instead, I want to address more fundamental questions: What is a technological revolution and what logic undergirds how successive technological revolutions have unfolded in Western societies? After answering this question, only then will I return to how we might fit the technologies of the so-called Fourth Industrial Revolution into a larger and more coherent context.

To begin developing an answer, I want to return to the year 1958 and to the community of Sevettijärvi, which is located 200 miles above the Arctic Circle in the northeast corner of Finland. The events that transpired there between 1958 and 1971 reveal with stark simplicity the core dynamics of change in an infrastructural technology unconfounded by the complexity of our here and now or by the prognostications of would-be futurists. In short, I shall deal in history whose interpretation, though always debatable, is somewhat more grounded than predicting the future. My telling of Sevettijärvi's tale draws heavily on the anthropologist Pertti Pelto's (1973) marvelous little book, *The Snowmobile Revolution: Technology and Social Change in the Arctic*. Pelto's ethnography is a little-known classic that all students of technology and work should read.[1]

As you might guess, in 1958 Sevettijärvi was relatively isolated from the then modern world. The nearest town was 120 miles away. The region was a land of snow and lakes, a harsh environment by most peoples' standards. Snow typically began falling in October and continued until May. On average the region experienced 210 days of snow cover each year, with an average depth of 2 feet.

---

[1] As of May 5, 2020, Google Scholar indicates that *The Snowmobile Revolution* has received 313 citations; such a pity for such an important book.

In 1958 most of the 335 inhabitants of Sevettijärvi were Skolt Lapps who had migrated there from their ancestral land to the east, which became part of the Soviet Union after World War II. Although the Russians had offered the Skolts the opportunity to stay in their homeland, they chose Finland instead. The Skolt's way of life revolved around reindeer herding. Each family had its own herd of deer. The deer provided food, clothing, and transportation.[2] Venison was sold to traders for cash, which the Skolts used to purchase flour, sugar, coffee, salt, and other staples.

Because of the vast distances between homes, the need to make long trips lasting several days to secure supplies, and the herding lifestyle, Skolts prized mobility. Roads in the region were poor and few. Skis and reindeer sleds were the primary means of transportation. Each family kept one or more geldings, which they used as mounts, and the Skolts also used deer as pack animals.

The Skolts maintained an open range: Grazing land was a free good. Any Skolt had the right to exploit the region's natural resources and herding required little starting capital. In fact, deer were routinely given to children as birthday, holiday, and wedding gifts. By the time children were ready to marry and start families, they already had accumulated a small herd of their own. For these reasons, there was no appreciable economic stratification in Skolt society, although some Skolts had herds that were larger than those of others.

In contrast to the reindeer of the Western Lapps, who inhabited Sweden and Norway, the Skolt's deer were tame, domesticated animals. "Reindeer tend to become habituated to the territories and/or migration patterns that they experience during their first year of life. Thus, the Skolts' reindeer did not engage in extensive migrations because the animals' ancestors from early times were not migratory" (Pelto, 1973: 32). Accordingly, even when allowed to graze freely during summer months, the Skolt's deer did not venture far from home.

The rhythm of Skolt life revolved around the reindeer cycle, as it had for thousands of years. In the summer the deer were allowed to

---

[2] Fish were also a staple in the Skolt diet.

roam freely on the tundra and in the forests, fattening themselves on grasses, leaves, mushrooms, and other vegetation. Finnish Lapland was divided into reindeer districts, each overseen by a reindeer association, a producers' cooperative responsible for staging roundups. After the first snowfall in October or November, the association hired herders who skied into the woods and across the tundra to gather deer and lead them to designated pasture areas. As the herding progressed, the assemblage of deer became larger and larger. Once it was apparent that most of the deer in the area had been assembled, often numbering in the thousands, the herders began to move the deer toward a communal corral to the south where a roundup would be held.

The trek to the corral was a stately procession, guided by a herder leading a gelding with a bell tied around its neck. Because the deer were relatively tame, the entire herd followed the lead gelding slowly in a long line that sometimes stretched on for a kilometer. Skiers traveled at the rear and alongside of the herd to discourage stragglers. After several days' journey, the deer began entering the corral. Once inside, the roundup began. Owners roped and separated their deer from the larger herd. The task was relatively easy because each family had its own brand, a unique pattern of notches cut into a deer's ear. Although the Skolts branded many calves in the spring before they allowed them to follow their mothers into the tundra, the Skolts could easily identify who owned calves that were born after the does' release because calves tended to remain near their mothers. The Skolts assumed that any calves congregating by a doe belonged to the doe's owner. Once a family assembled its herd, they branded unbranded calves, slaughtered deer for meat, and bartered the meat to secure supplies from traders who frequented the roundups.

At the conclusion of the roundup and after much feasting, families led their deer back to grazing areas near their homes. During the winter, herders tended their deer. They butchered deer for meat as needed and castrated bulls to create geldings for draft animals. If supplies ran short, the men slaughtered a few deer for their meat and then trekked with sleds to trading posts across the border with Norway, where they traded meat for supplies. During the spring, owners captured their pregnant does and tethered them near the house so they could give birth in an environment safe from predators. After birthing, owners branded the calves. By June, the owners had released the herd into the tundra and the cycle began again.

This was the state of affairs until 1962, when Canadians brought the first snowmobile to Skolt Lapland. Because the Skolts prized mobility, they immediately recognized the snowmobile's advantages. With a snowmobile a herder could complete a trip to Norway for supplies in six hours instead of four days. By 1971, all but twelve of the seventy families in the Sevettijärvi region owned a snowmobile, and several owned two. As evidence of the snowmobile's rapid adoption, one year after the snowmobile's appearance, the reindeer association decided to experiment with snowmobiles during roundups. Pelto's story is about how the snowmobile transformed Skolt society in less than a decade. The key to the transformation lay in the snowmobile's effect on reindeer herding.

Compared to herders on skis, men mounted on snowmobiles could complete a roundup in a matter of days rather than weeks, and there were economic incentives to do so quickly, because men on snowmobiles cost the association more than men on skis. With snowmobiles the Skolts also found there was less need to keep a winter herd nearby. They could use a snowmobile to track down deer easily when they needed food, and with a snowmobile they no longer needed deer to pull sleds or geldings to ride. But most importantly, snowmobiling began to change the ecological relationship between the deer and the Skolts.

In contrast to traditional practices, snowmobiling was noisy and frantic. The machines frightened and scattered the deer. To escape, the deer sought out increasingly inaccessible territory. Consequently, herders had a more difficult time finding their deer because snowmobiles could not maneuver over rocky, craggy land. Between 1962 and 1968 the number of deer gathered during roundups fell by 67 percent.

During mechanized roundups the deer kept running. As a result, an increasing number of unbranded calves became separated from their mothers, which made it difficult for owners to identify their calves. In the first eight years of snowmobiling the number of unidentified calves increased by 79 percent. If families could not identify their calves, the only way to increase the size of their herd was to buy calves at an auction held at the end of the roundup. The association's rules stipulated that calves would only be auctioned in lots of twenty to thirty. As a result, only more wealthy Skolts could afford to buy calves to increase the size of their herd. As this fact suggests, snowmobiling changed the economics of reindeer herding. Between 1962 and 1969, the cost of herding increased by 300 percent. The cost of staging the roundup increased by 500 percent, in large part because of the cost of running snowmobiles and the greater wages paid to snowmobilers. The increasing costs put pressure on the reindeer association to hold even speedier roundups, which exacerbated the cycle of decline.

By 1971, less than ten years after the first snowmobile arrived, most families owned at least one snowmobile, but the cost of herding had become so great that 73 percent of the families dropped out of herding altogether. The increasing centrality of cash to the economy led to economic stratification. Only families with money could accumulate large heads by buying calves at the auctions. Since larger herds begot even larger herds, the rich became richer. By 1971 three families owned 31 percent of the reindeer in the region. The median number of reindeer owned per family fell from fifty-two to twelve. As Skolt men dropped out of reindeer herding, they either became snowmobile mechanics, few of whom were required, or they resorted to the Finnish welfare system. The increasing importance of cash to the economy led to higher rates of out-migration among young Skolts, especially young women, to larger Finnish cities. Educational and occupational aspirations followed socioeconomic lines. Children of those families who were better off disproportionately took advantage of opportunities to "better themselves" by leaving the region. Thus, in less than ten years the snowmobile had reorganized Skolt society by shifting the foundations of economic, communal, and family life.

The case of snowmobiles in Skolt Lapland is instructive because it underscores with force and elegance a rare but important type of technological change. Broadly speaking, technological change comes in two types: *substitutional* and *infrastructural*. Most technological change is substitutional: A more efficient or effective new technology replaces an older technology used for the same purpose. Examples of substitutional technologies are legion: the use of pens instead of pencils, the replacement of picks for jackhammers, the shift from prop engines to jet engines, the replacement of vinyl records by CDs and then by MP3s.

Technological substitutes usually reduce the costs of production and generate considerable profits. But—and this is important—their effects tend to be confined to isolated areas of human activity because most substitutional technologies tend to be peripheral to a society's system of production. With microwave ovens we can cook faster, but aside from changing how we cook (and perhaps what we eat) microwaves have not substantially altered the way we live. CDs are more durable than vinyl records and arguably produce better sound, but they did not change the act of listening to music.

Substitution is the most familiar sort of technological change, and it is the kind of change that our institutions are set up to handle. Businesses, for example, are always looking for technologies that allow them to produce goods or services of greater quality, in greater quantity, and at lower cost. Consumers are attracted to new tools that allow them to make household chores easier or that improve the quality of their activities such as watching TV at ever high resolutions on increasingly larger screens with increasingly smaller pixels.

Infrastructural technological change is entirely different. Infrastructural technologies are the *small set* of technologies that form the cornerstone of a society's system of production during a particular historical era. The Skolts' infrastructural technologies were their traditional herding techniques. In industrial societies, until most recently, the key infrastructural technologies were electric power, the electric motor, the telephone, and the internal combustion engine, all of which came into being in the late nineteenth century and formed the basis of what scholars call the Second Industrial Revolution.[3]

---

[3] I shall use "Industrial" and "Technological" revolution interchangeably throughout the remainder of this chapter.

Because infrastructural technologies lie at the core of a society's economy, they eventually underwrite almost every other aspect of daily life. The Skolt's traditional herding practices not only governed the round of life; they shaped how the Skolts spent their time, where they lived, the nature of their relations with others, and how they adapted to the region's ecology. A reshaping of society by infrastructural technologies also occurred during the Second Industrial Revolution. Electricity, the electric motor, the internal combustion engine, and the telephone set the stage for the emergence of large corporations, a trend that began with the railroads a few decades earlier (Chandler, 1977). Infrastructural technologies and the network of subsidiary technologies that augmented them brought fundamental changes in the occupational structure, for example, the emergence or growth of such occupations as machinists, electricians, automobile mechanics, tool and die makers, electrical engineers, mechanical engineers, clerks, and even management itself. Because electricity freed factories from being located near natural sources of power, the rise of factories led to greater urbanization as towns grew up around the factories and as people flocked to towns from rural areas to take factory jobs.[4] Increasing urbanization shrank the extended families characteristic of rural life into nuclear families as people migrated to cities leaving their kin behind. With the expansion of cities and the diversity of the people they attracted, urban life gradually became more secularized. In short, as illustrated in Figure 1.1, when societies experience changes in their technical infrastructures, one should expect ramifications eventually to extend throughout most institutional sectors of the society, thus altering the contours of people's lives. This is precisely what happened with the Skolts. In essence, snowmobiles represented a wholesale shift to mechanized sources of power, eventually altering social networks and the structure of families. What sets the Skolts apart from us is how clearly and quickly the transformation occurred.

---

[4] To be accurate, steam engines and the railroads, properly associated with the First Industrial Revolution, played a significant role in the emergence of factories and the growth of urban areas, but these trends were greatly enhanced by the technologies of the Second Industrial Revolution.

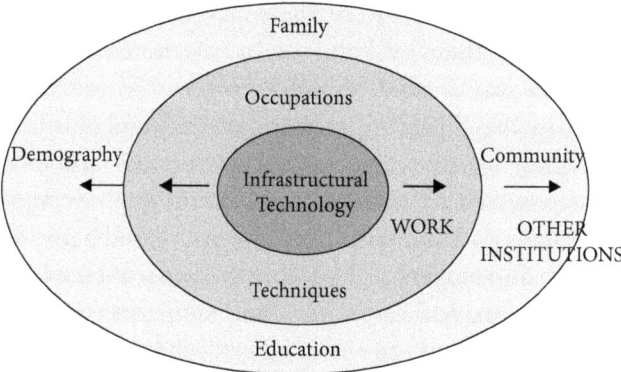

Figure 1.1. Infrastructural Technologies Occasion Change in Work That Ramify Across the Institutional Sectors of Society

Because infrastructural technologies undergird systems of production, they begin changing societies and a people's way of life by transforming the nature of work. The change occurs on two fronts: what people do for a living and how people do what they do. Note that after adopting and integrating snowmobiles into reindeer herding, most Skolts stopped being herders. A few became mechanics, others immigrated to cities and took up a variety of jobs, and the remainder ceased working altogether. The Second Industrial Revolution had the same effect: In the first half of the twentieth century farmers became factory hands, housewives and young women became clerks and secretaries (which had previously been men's work), foremen and male clerks became managers, livery operators became cab drivers, and blacksmiths went out of business. The list could go on and on. In other words, changes in infrastructural technologies gradually alter the division of labor in society.

By changing how people do what they do, infrastructural technologies also shift modes and means of production. In the case of the Skolts, the shift was simple: They moved from a manual to a mechanized mode of herding. The same shift occurred as a result of the infrastructural technologies of the Second Industrial Revolution. The mode of production changed from handicraft production toward the machine manufacturing of goods previously produced by artisans. Furniture was produced by lathes and other machine tools driven by electric motors instead of the hand tools used by woodworkers. Metal parts were made on machine tools powered by electricity rather than by hand or by machines driven by belts whose motion was created by water power or steam engines. The change led to the flowering of standardized and interchangeable parts. Farming became mechanized as tractors and other farm equipment powered by internal combustion engines replaced the labor of humans and animals. Clerical work ceased being done with pens, pencils, and ledgers and became the work of people operating typewriters, mechanical calculators, and tabulating machines. With incandescent bulbs and the electrical grid, shift work spread. It was now possible for some people to work during the day, while others worked through the night, thereby keeping factories running twenty-four hours a day.

The problem with infrastructural technological change is that people often view these technologies as another instance of substitutional change. In fact, infrastructural technologies do usually have substitutional implications. For example, it is not difficult to imagine a Skolt herder framing the snowmobile as simply a faster sled or gelding. They did not see the snowmobile as a technology that would challenge their relationship with reindeer. Similarly, electric lighting was a more effective and efficient substitute for gas lighting, candles, or oil and kerosene lanterns. It was not seen as a way of driving away the night and extending the workday. Our tendency to view infrastructural technologies initially as substitutional is understandable. When infrastructural technologies first appear, it is difficult to imagine how they will spawn a network of related technologies, applications, and forms of organizing that have yet to come into being. The misapprehension arises because we tend to focus on what Sproull and Kiesler (1991) call a technology's "first order effects."

First-order effects tend to be primarily economic and cast in utilitarian terms: They reduce costs, increase quality, or enhance the doing of some task or the making of some good. It is by appealing to first-order effects that vendors typically sell technologies to users. They are successful at doing so because first-order effects tend to occur relatively quickly. Perhaps most importantly, first-order effects are relatively predictable. Snowmobiles and automobiles do get you places more quickly than reindeer sleds or horses and wagons. A computer programmed with the right software does perform calculations faster than one can do by hand or even with an adding machine or a tabulator, and it is likely to make fewer mistakes. Compared to a gunsmith, electric-powered machine tools produce more guns more quickly and more identically.

In contrast to first-order effects, second-order effects are sneaky. They are hard to foresee, and they are rarely utilitarian. Electricity and electric lighting gradually changed people's diurnal patterns. Internal combustion engines and the vehicles they propelled allowed people to become more mobile and less tied to place (Flink, 1975). Telephones made people less socially isolated, as farmer's wives discovered in the early days of the telephone's diffusion (Fischer 1992). Electric washers and driers not only made it easier to wash clothes; they eventually changed our standards of cleanliness and even what constitutes a suitably varied wardrobe (Cowen, 1985). In short, second-order effects tend to be sociocultural.

With the idea of an infrastructural technology firmly in mind, we can offer a reasonable definition of an industrial or technological revolution. Such so-called revolutions entail a fundamental change in the technological infrastructure of a society and, hence, in the social organization of a society's productive activity. Historians have a difficult time untangling the relative importance of the two. Rather than debate the issue, let us accept the idea that the two are mutually constitutive or reinforcing: a theme recently sounded in micro-level studies of technological change under the banner of socio-materiality (Orlikowski and Scott, 2008).

Instead, let us turn our attention to several myths about technological revolutions that are worth dispelling because they appear so frequently in popular discourse such as that articulated by advocates of a Fourth Industrial Revolution. The first myth is that a new infrastructural technology "causes" a revolution. It is reasonable to argue that technological revolutions rest on a handful of core technologies, although candidates for what those core technologies might be are usually difficult to see without hindsight. Nevertheless, the emergence of a core technology is not sufficient. When one closely examines what happened during the eras when technological revolutions are said to have occurred, one sees what Schumpeter (1934), the great economist of innovation and entrepreneurship, called a "swarming" of innovations. A similar case is made by students of economic long waves; the 50–80 year cycles that are often associated with what in retrospect we call a technological revolution (Kondratieff, 1935; Rostow, 1980; van Duijn, 1983; Coombs, 1984).[5]

Consider the technological innovations of the late nineteenth and early twentieth centuries. In addition to the electrical grid, the electric motor, the internal combustion engine, and the telephone, there were a host of other new technologies that emerged either to support and produce the new core technology or to build on the core technologies' capabilities. We might call such sets of adjunct technologies "technological systems" that occasioned change in many realms of life (Hughes, 1994). A short, incomplete, but illustrative list of technologies that swarmed during the Second Industrial Revolution would include light bulbs, electric irons, automatic washing machines, the standardization of alternating current, vulcanized rubber, electric generators both large and small, transformers, vacuum cleaners, macadamized roads, and machine tools. In addition to these technologies, social innovations, such as assembly lines, standard time zones, and bureaucratic corporations were crucial for organizing the use, control, and expansion of the technological systems built around and on top of the infrastructural technologies.

Precisely why technological innovations tend to swarm after the emergence of an infrastructural technology is difficult to explain (see Freeman, 1984). There are probably many entwined reasons. Infrastructural technologies, like basic discoveries in science, open related possibilities. These possibilities tend to attract capital and entrepreneurs who see ways of improving, building on, or taking advantage of the new core technologies. Some scholars, such as

---

[5] Students of long waves contend that over the last 200 years Western economies have experienced four broad cycles of expansion and contraction, each with a period of approximately fifty years (see Barley and Kunda, 1992).

Beniger (1986), argue that infrastructural technologies pose subsidiary problems that have to be worked out before the core technology can be adequately used or deployed. This was certainly how the electrical grid was built over time: Electricity's utility (no pun intended) required a host of related technologies, both technical and social, such as generators, transformers, wires with appropriate resistance and shielding, power plants, transmission lines, regulatory bodies, utility companies, and much political maneuvering (Hughes, 1983; Granovetter and McGuire, 1998).

The second myth might be called the fallacy of materialistic determinism.[6] A materialistic and deterministic account focuses on technologies as direct causes of social change. The technology arrives on the scene and change unfolds with a kind of inevitability, often called progress. Deterministic arguments about what new technologies will let us do are rampant. This is the bread-and-butter discourse of Silicon Valley's technological evangelists, and it is the discourse of advocates of the Fourth Industrial Revolution. But we should not hang the specter of determinism solely on our contemporaries. Determinism was just as rampant in the nineteenth century, when the concept of progress gripped the popular imagination (Smith, 1994). Thus, contemporaries heard, and we still sometimes hear, such statements as "The railroads allowed the taming of the American West." Of course, Native Americans had a different perspective on the matter.

The point to make is that infrastructural technologies are necessary for industrial revolutions, but they are not sufficient. First, technologies can be used in a variety of ways, and people

---

[6] As I have discussed more fully elsewhere with respect to technologically driven change (Leonardi and Barley, 2008: 160):

Determinism holds that our actions are caused by technological, cultural and other forces prior to, external to, and independent of our behavior. Voluntarism takes the opposite stance, arguing that humans have agency (what philosophers call "free will") and can shape their environments to achieve their interests and goals. To be a materialist is to hold that human action stems from physical causes and contexts such as geography, biology, climate, and technology. Conversely, idealists argue that ideas, norms, values, ideologies and beliefs (what most of us call the social) drive human action. Although the distinction between determinism and voluntarism is orthogonal to the distinction between materialism and idealism, social scientists frequently write as if materialism implies determinism and idealism implies voluntarism. This is simply not the case.

(especially powerful people and policymakers) make choices about those uses. For example, there is nothing inherent in the snowmobile itself that "made" the Skolts decide to use the snowmobile for reindeer herding. They could have viewed it as simply a means of travel or even a recreational activity, as most Canadians did. Similarly, there was nothing inherent in the electric motor that foreordained the rise of assembly lines. The combination of motors and conveyor belts to create the assembly line was an idea promulgated by Henry Ford and his managers to achieve their own objectives. In short, individual choices, corporate strategies, and public policy are important for how a technology is deployed and, hence, for the effects that it may or may not have.

Ideas also matter. The concept of efficiency most closely associated with Frederick Taylor's *Scientific Management* (1911) and the ideas that underwrite the form of organization we call bureaucracy were critical to how the Second Industrial Revolution evolved (Weber 1922/1968). In fact, efficiency was every bit as important a trope in the late nineteenth and early twentieth centuries as the concept of progress (Haber 1964). So enthralled were Americans with the idea of efficiency that a series of events between 1910 and 1912 transformed scientific management almost overnight into what seems to have been the first American business fad.

In 1910, the Eastern Railroad requested a rate increase from the Interstate Commerce Commission (ICC). The request caused widespread anger among the middle class and among industrialists, who felt that rates were already too high. Louis Brandeis agreed to represent a group of industrialists who challenged the increase before the ICC.[7] Brandeis argued that had the railroad been managed more efficiently, it could have met its costs without raising prices (Nelson, 1980). To support his claim, Brandeis solicited testimony from key Taylorites. Their testimonies not only became the centerpiece of the hearings and created a national audience for Taylor's ideas, but the term "scientific management" was apparently coined during the hearings (Haber, 1964).

---

[7] In 1916 President Woodrow Wilson nominated Brandeis to the Supreme Court of the United States. The Senate confirmed his nomination.

Taylor and his protégés used the Eastern rate case to further popularize their views. Immediately afterward, Taylor (1911) published *The Principles of Scientific Management,* which became a bestseller. Harrington Emerson, a self-proclaimed spokesman for the movement, published two even more popular books lauding the benefits of efficiency (Emerson, 1912, 1914). The Society to Promote the Science of Management (later called the Taylor Society) and Emerson's Efficiency Society were both founded in 1911. These and other developments occasioned a public mania known among historians as the "efficiency craze." The rhetoric of efficiency became so popular in America that in 1914 an "efficiency exposition" was held in New York City, with Taylor as the keynote speaker. The exposition drew a crowd estimated at 69,000 (Haber, 1964: 61). There is little doubt that the notion of efficiency shaped how the technologies of the Second Industrial Revolution were deployed to reconfigure the nature of work, not only in factories but in offices as well (Yates, 1993).

Finally, the term "revolution" is a misconception. "Revolution" invokes a discrete disjuncture in technological, economic, and social history. It is a term usually associated with politics and implies the overthrow of a former social system in a relatively short period of time, as in the American Revolution or the Russian Revolution. Given the swarming of technologies that mark so-called technological revolutions and the relatively long time it takes to put technological systems in place, it is perhaps more accurate to talk about technological evolutions that are punctuated by phases of intensification. Moreover, careful examination of the historical record frequently suggests that people were experimenting with technologies and techniques that foreshadowed the developments usually associated with the advent of a new infrastructure. Beniger (1986) is particularly persuasive on this point. He argues that what some scholars (and I will) call the control revolution has been working itself out for at least 200 years.

To put the matter of technological revolutions into perspective regardless of how they occur, how long they take, or how unevenly they affect particular industries, I shall draw on another much-overlooked work of scholarship: William Faunce's "Automation and the Division of Labor" published in *Social Problems* in 1965. It is important to recognize that Faunce wrote when integrated circuits and semiconductors were still in their infancy. Although Fairchild Semiconductor had been founded eight years earlier, Intel would not be established for another three years. Yet Faunce had insight into what computers would ultimately do based on studies of continuous flow plants in which control systems were already being automated by electromechanical technologies (Blauner, 1964).

Faunce began by observing that all production systems are composed of four components:

> The first of these is *power technology* dealing with the sources of energy used in production. A second is *process technology* [which I shall call a conversion technology], which refers to the tools and techniques used in the actual operations performed on raw materials. A third is a *materials handling technology* dealing with the transfer of materials between processing operations. The final component of production technology is *control* or the regulation of the quality and quantity of output. Any production process involving the conversion of raw materials into finished products can be separated into its power, processing, materials handling and control components.   (Faunce, 1965: 150, italics in original)

Furthermore, Faunce noted that any of these components could be performed *animately*, that is, by people or animals. Or they could be performed *inanimately* by machines. Combining Faunce's notion of animate and inanimate means with his four components of production yields a framework for interpreting the broad history of infrastructural revolutions as displayed in Figure 1.2. As the figure suggests, the history of technological change over the last several centuries has been a slow push toward the mechanization of all four components of a production system.

Prior to the First Industrial Revolution, when all products were made by hand, by either craftsmen or users (such as farmers), humans or animals performed all components of production systems. Animals were used primarily as a source of power and to move raw materials and goods from one point to another. At best, hand tools augmented human labor. The only notable exceptions

Components of a Production System

Period	Power	Conversion	Transfer	Control
Handicraft Era	Animate	Animate	Animate	Animate
First Industrial Revolution	Inanimate	Mixed	Animate	Animate
Second Industrial Revolution	Inanimate	Inanimate	Mixed	Animate
Control Revolution	Inanimate	Inanimate	Inanimate	Inanimate

Figure 1.2. Faunce's Theory of Technological Revolutions

were the use of wind and water mills to grind grain and sailing ships that also used wind power.

The First Industrial Revolution (roughly spanning the late eighteenth through the mid- nineteenth century) entailed a shift to inanimate sources of power: initially water power and then energy supplied by steam engines. Steam engines had a variety of applications from pumping water from mines to creating energy to drive production machinery through systems of cams, gears, pulleys, and belts. The key to the steam engine, whose invention is often misattributed to James Watt, was the transformation of reciprocating motion into rotary motion.[8] Rotary motion was critical for suppling kinetic energy to line shaft drives: a system of camshafts and belts that drove devices such as looms and spinning wheels. Thus, the inanimate power sources of the First Industrial Revolution allowed preliminary experimentation with inanimate conversion technologies. For this reason, in Figure 1.2, I used the word "mixed" in the third column of the second row to represent a mixture of animate and inanimate conversion components. However, it is worth remembering that the automation of conversion was rudimentary, initially resting on water power, and was deployed primarily for manufacturing textiles, first in England and much later the United States.

---

[8] Thomas Newcomen built a functional and much-used steam engine for pumping water from mines which he patented in 1781 nearly seventy years before Watt's rotary motion engine (Basalla, 1988). Steam engines underwent continual improvement until at least the 1930s.

During Second Industrial Revolution (the late nineteenth to the early decades of the twentieth century), the electric grid, the electric motor, and the internal combustion engine completed the shift to inanimate conversion technologies. Machine tools and other manufacturing equipment were now equipped with their own motors, eliminating the necessity of line shaft systems for activating machines, which, in turn, opened up the possibility of mass manufacturing on a larger scale. Furthermore, electric motors enabled the mechanization of large numbers of conversion tasks not associated with factories, from washing clothes to cleaning floors, to cutting and routing wood, to pumping hot water through homes. Electric motors and internal combustion engines also allowed experimentation with inanimate transfer systems. For example, automobiles on Ford's early assembly lines were moved by tow ropes activated by a capstan. What these early assembly lines lacked was sophisticated control technologies that regulated not only the operation of machine tools, but also the speed of conveyors that made transfer technologies possible. Many production facilities continued to move finished and unfinished goods by hand. Accordingly, Figure 1.2 indicates that transfer technologies associated with the Second Industrial Revolution were mixed.

With the advent of servomechanisms and ultimately the spread of semiconductors, microprocessors, and computers, production entered a period that some have called the "Control Revolution" (Beniger 1986). I prefer this term to the "Third Industrial Revolution," because it is more descriptive and signifies a more evolutionary than a revolutionary process. Control technologies based on servomechanisms were initially widely applied in continuous-flow plants that made paper, chemicals, and gasoline (Woodward 1958; Blauner 1964). In such plants, operators could monitor production processes from control rooms using gauges and dials and, in response, alter temperatures, pressures, and the speed of the conversion process. As early as the 1960s, computer control was applied to machine tools, first through numerical control (NC), which involved programs punched on paper tapes that guided cutting paths in predetermined sequences and at high tolerance (Noble 1984). With the advent of the microcomputer, in the early 1980s computers were mounted on the machine tools themselves (this was called computerized numerical control or CNC). With the development and refinement of robots, numerically controlled machine tools, computer-controlled transfer systems, and computer-aided design and computer-aided manufacturing systems, by the late 1980s it became possible to run entire plants with only a few technicians. In 1985 Bylinsky and Moore (1985: 287–8) wrote of a factory they observed in Japan (which was ahead of the U.S. and Europe in factory automation):

> In Fanuc Ltd's cavernous, bumblebee-yellow buildings...automatic machining centers and robots typically toil unattended though the night, with only subdued blue warning lights flashing as unmanned delivery carts move like ghostly messengers through the eerie semi-darkness. This plant...makes parts for robots and machine tools (which are assembled manually, however). The machining operation...is supervised at night by a single controller, who watches the machines on close-circuit TV. If something goes wrong, he can shut down that particular part of the operation and reroute the work around it...The total cost of the plant was about $32 million, including the cost of 30 machining cells, which consist of computer-controlled machine tools loaded and unloaded by robots, along with materials-handling robots, monitors and a programmable controller to orchestrate the operation. Fanuc estimates that it probably would have needed ten times the capital investment for the same output with conventional manufacturing. It also would have needed ten times its labor force of about 100. In this plant one employee supervised ten machining cells; the others act as maintenance men and perform assembly.

Further developments in computer-aided design, computer-aided manufacturing, and the emergence of the Internet have made it possible to design a part in one place, send the design via the Internet to a factory somewhere else, and then have machine tools make the part without human intervention. In essence, the control of production processes is gradually shifting from human operators and mid-level managers to hierarchies of computers overseen by a small number of technicians (Zuboff, 1988; Vallas, 2001).[9] Faunce foresaw this possibility and argued in 1965 that progression in automation would make not only transfer technologies, but control technologies, inanimate. At this point humans would presumably be peripheral to, if not "removed from the loop," as engineers would put it.

When considering the control revolution, it would be a mistake to assume that entirely inanimate control systems are confined to manufacturing, on which Faunce largely focused. We have moved to a point where inanimate technologies can manage retail operations and the delivery of services. Several well-known examples will serve the purpose of making the point. Through Amazon and similar websites, consumers can now easily order and pay for goods via a sophisticated system of algorithms and computational devices that are activated by keystrokes from their home computers or cellphones. Once Amazon's computers receive the order, they instruct robots to retrieve the ordered goods from shelving in a warehouse and then move them to a mailing station, where a human picks the item from the shelf and prepares it for shipping.[10]

When you use Google to search the Web for information, you are activating sophisticated algorithms to return information on your requests. The computers keep track of your requests and eventually serve you not only the information you requested but ads tailored to your interests and preferences. This is just one application of what computer scientists refer to as big data: the ability to make inferences about individuals and situations based on patterns contained in large quantities of data. The smart grid is another example. By imbedding microprocessors in the electrical grid, it will be possible to monitor more cheaply the use of electricity, to alter the amount of electricity flowing from one point to another, and,

---

[9] For example, you can view a fully automated facility for manufacturing contact lenses at https://www.youtube.com/watch?v=r19M9PZbK58, accessed March 25, 2020. BMW's automated production system can be viewed at https://www.youtube.com/watch?v=fgyDo_1d390, accessed March 25, 2020.

[10] To see how Amazon uses automation, see https://www.youtube.com/watch?v=ezyfyZ3LTE4, accessed March 25, 2020, and https://www.youtube.com/watch?v=wC4vITSVXoA, accessed March 25, 2020.

ultimately, to charge a variety of rates for electricity depending upon when the electricity is consumed. Meter readers may become an occupation of the past.

Self-driving cars, which have begun to attract so much attention, are essentially inanimate control systems that rely not only on sensors and computers embedded in the car but on GPS and other data fed to the vehicle from a microelectronic infrastructure. Microsoft and other companies are even experimenting with making inanimate humans. Microsoft has been developing an office assistant which can respond to humans much like another human. The machine currently displays an avatar on a computer screen outside a manager's office. Language recognition programs and other information processing technologies allow the inanimate human to inform an animate human whether a manager is currently available, when he or she will be available, schedule an appointment, and leave the animate human's messages for the manager.[11] Although advocates of the Fourth Industrial Revolution suggest that this is a quantum change, such technologies are better understood as an evolutionary outgrowth of the control revolution rooted in digitization that began in earnest during the 1980s based on the improvements in semiconductors, microchips, sensors, and machine learning whose early prototypes were developed in the 1960s. Today, digital control technologies are, to use Schumpeter's term, "swarming."

They key question is what happens to our society, culture, and way of life when an increasing number of control systems become inanimate. Faunce argued that each era was accompanied by a dominant form of organizing work, although he did not elaborate on what they were. It is fair to say that the primary forms for organizing work during the handicraft era were slavery, guilds, apprenticeships, and small shops. During the First and Second Industrial Revolutions we saw the emergence of factories and, later, large corporations and bureaucracies. Employment relations moved from outsourcing, the putting-out system, and the drive system to what students of industrial relations now call the bureaucratic or

---

[11] To see the AI personal assistant in action, view Eric Horvitz's TED talk at https://www.youtube.com/watch?v=dpoVh9xwdD4, accessed December 29, 2019.

traditional employment system marked by unions, internal labor markets, labor laws that protected workers, and the promise of job security in return for loyalty and effort. As we move further into the control revolution, we have witnessed the decline of unions, a return to contracting and outsourcing, the unraveling of job security, the increasing allocation of employment to people on the basis of specific tasks or skills, and high-velocity labor markets in which people change what they do and who they work for with greater rapidity. Some have even argued that network structures are replacing bureaucracies and corporations as a form of organizing work (Powell, 1990; Davis, 2016).

In the past people, feared that waves of automation spawned during the First and Second Industrial Revolutions would lead to widespread unemployment. Fortunately for our forefathers, this did not occur, because the demand for goods and services increased more rapidly than automation displaced labor, which therefore allowed the continued employment of individuals and even the expansion of employment. The question to contemplate is: Were we simply lucky in the past? During the First Industrial Revolution the opening of new markets, particularly in the Americas and elsewhere, enabled those who automated their conversion processes to sell their goods to larger numbers of consumers at reduced prices, which thereby increased rather than decreased employment. During the Second Industrial Revolution we experienced World War I and more importantly World War II, which decimated the productive capacity of many of the world's most productive economies. The end of the World War II created a significant pent-up demand for goods, particularly in Europe and Asia, but also in the United States and Canada—which were well situated to supply the demand, because their industrial infrastructures escaped unscathed. Domestic markets also expanded as wages rose. Can we expect the same today? The answer is unclear. With the shift to inanimate control systems, production of all types, from goods to services, may well require fewer people, albeit perhaps with greater technical skills (although in Amazon's case that does not appear to be true). Furthermore, the emergence of inanimate control systems has occurred simultaneously with the tendency for firms in North

America and Europe to outsource jobs to parts of the world where labor is much cheaper.

Technologies that increasingly control both themselves and other machines create challenges we have not faced in the past, primarily because control was the last unassailed bastion of employment. Consider self-driving vehicles: What will truck drivers do when computational devices can drive trucks and deliver goods? It is worth noting that truck, delivery, and tractor drivers are the largest employers of men in thirty-one American states.[12] Consider medicine: What will radiologists do when computational devices can interpret medical images? Similarly, what will surgeons do when surgical robots no longer need to be guided by human hands but can operate using sophisticated medical images and machine-learning algorithms? Finally consider law: What will lawyers do when artificial intelligence can produce wills, set up trusts, or file for divorce? As Levy and Murnane (2005) have argued, perhaps the only jobs that remain secure are those of the trades (plumbing, masonry, carpentry) and personal service occupations which are difficult to either automate or outsource.

I do not believe it is reasonable to sound the alarm of the Luddites, at least not yet. But I do mean to spur your consideration of what our societies will be like when all components of an array of production systems ranging from the manufacturing of goods to the provision of information and knowledge can be augmented and perhaps performed by the inanimate. The answer to this question burrows to the core of our institutions, from education to the family to the government (see Bailey and Barley, forthcoming). Chapter 3 takes up the issue of intelligent machines' implications for work and employment. For now, suffice it to say that it is time for us to think through what kinds of research we will need to inform our thinking and our actions. The advocates of the Fourth Industrial Revolution may be right when they say big changes are coming, but my point is that they have been coming for a while, just as in previous technological revolutions. I suspect that the most

---

[12] See http://www.npr.org/sections/money/2015/02/05/382664837/map-the-most-common-job-in-every-state, accessed March 25, 2020.

difficult challenges are not likely to be simply technological in nature; they will most definitely be social, economic and political. The danger is that we mistake what commentators say may be already happening for the inevitable and, hence, delay exploration and debate about how to deal with the challenges we face and the choices we will need to make to adapt to a much-transformed society. Before returning to the question of how intelligent machines may occasion changes in work and employment, let us turn to the question of how technologies change work systems and organizations.

2
How Do Technologies Change Organizations?

Chapter 1 focused on large scale sociotechnical developments that span decades and define eras. As Thomas Hughes (1994) noted in his essay, "Technological Momentum," when you view technological change from a macrosocial perspective, it is easier and perhaps even reasonable to speak more deterministically, because from that vantage point you can see dominant trends, particularly when analyzing the development of technical systems and infrastructures over the *longue durée*. This chapter ratchets the resolution down by moving away from technical infrastructures and broad social changes to consider specific technologies in specific workplaces used by members of particular occupations at particular times and places. As an ethnographer of technology and work, I am most comfortable with this is the level of analysis, and it is where I have spent most of my career. As Hughes (1994) noted, as one gets closer to a specific technology's development and its implications for particular places at particular times, deterministic stories tend to evaporate into tales of social construction, contingencies, and the consequences of actions that might have otherwise been.

The first paper on technology I wrote challenged technological determinism at the level of the organization (Barley 1986), and since then I have written extensively about the social construction of the use of technologies (Barley 1988a; Barley, 1988b; Barley, 1988c; Barley and Tolbert, 1997; Orlikowski and Barley, 2001; Barley, Meyerson and Grodal, 2011). Indeed, the 1980s and 1990s witnessed a movement among students of technology and work toward more constructionist stances, with Wanda Orlikowski being

Work and Technological Change. Stephen R. Barley, Oxford University Press (2020). © Stephen R. Barley.
DOI: 10.1093/oso/9780198795209.003.0002

one the most influential advocates of the constructionist stance.[1] However, rejecting determinism does not mean rejecting materialism. Whereas determinism holds that our actions are *caused directly by* technological, biological, ideological, or other forces, materialism holds that our actions *are affected or shaped by physical contexts* such as geography, biology, climate, or technology that cannot be avoided. For example, you cannot walk through a wall, but you might be able to walk around it, climb over it, tunnel beneath it, or knock it down. Accordingly, over the past few years scholars have been thinking about how social constructionists might come to terms with the reality of material constraints and affordances and how they become woven into or entwined with human action (Orlikowski 2007; Orlikowski and Scott 2008; Leonardi and Barley 2008, 2010; Bailey, Leonardi, and Barley 2012; Barrett, Oborn, and Orlikowski, 2016). As Orlikowski (2007) argued, the move to social construction led scholars to place too much emphasis on human actions and decisions, while ignoring the realities of the physical and material. Somehow, we need to be able to combine the two.

This chapter recounts my observations of how two technologies altered the organizations that adopted them. After the telling, it turns to the issue of how to conceptualize social construction in a world of material constraints and affordances. My primary goal is to use studies I have done over my career to provide an answer to a question of significance to students of technology and organizing as well as those who run and work in organizations: How do technologies change organizations?

After nearly forty years of studying work, technology, and organizing, I have concluded there is only one certainty about technological change: You almost never get *only* what you expect and sometimes you do not even get that. However, something usually happens. Even if it is hard to predict whether and how a new technology will alter an organization, I submit that any significant organizational change will unfold according to an identifiable process. My view is that technologically occasioned organizational change occurs as a

---

[1] See Leonardi and Barley 2010 for a review of constructionist approaches to technological change in the workplace.

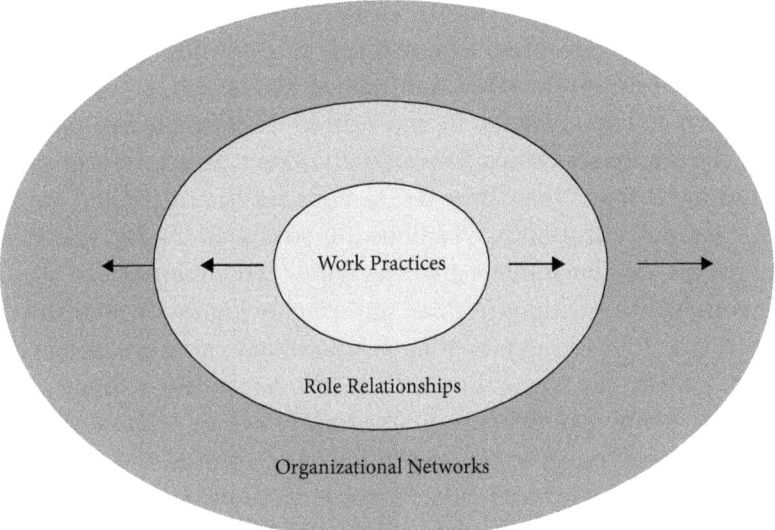

Figure 2.1. How Technologies Change Organizations

series of interlinked reverberations across levels of analysis that
eventually alter the organization's structure and culture. If a new
technology is going to do anything to an organization, it will start
by changing work practices: what people do, how they do it, or
both. If change stops here, a technology's effect on the organization
will be relatively limited, although it may be quite significant for
the people who use it. If a technology only changes work practices,
it will change neither how the organization is structured nor the
organization's culture or subcultures. However, if shifts in what
people do and how they do it affect *with whom they interact* or *how
they interact* with others, then the technology's reverberations cross
a critical boundary. They now begin to alter roles and role relationships. When roles and role relationships change, significant shifts
are likely to occur in an organization's division of labor and its
structure. The reasons are simple: (1) role relationships define social
networks (White, Boorman, and Breiger, 1976) and (2) social networks *are* an organization's structure (Krackhardt and Hanson,
1993). Roles, role relationships, and networks are, therefore, the
theoretical concepts with which I am concerned. Figure 2.1 depicts
the logic of the process.

Three concepts are critical to a role-based approach to technological change: position, role, and role relationships. By position I simply mean an identified and named status, job, or occupation within a division of labor. By role I mean what people in a position do. By role relationships I mean with whom people in a position must interact and how those interactions are structured and unfold in everyday encounters. Positions are structural. Roles, role relationships, and encounters are behavioral.[2] Technologies can attach themselves to positions in three ways: technologies can augment or otherwise transform an existing position, they can eliminate a position, or they can trigger the creation of a new position. Sometimes, they do some combination of all three. However, technologies do not do these things automatically and deterministically. In organizations, these outcomes either involve decisions made by those in power or emerge through time in an evolving process of concerted adaptation that involves human action. Regardless, given a position, it is roles, relations, and encounters that become the primary foci of analytic concern. This stance reflects my grounding in American Pragmatism as it was developed by the Chicago School of Sociology associated with Park and Burgess (1921); Mead (1934) and, especially, Hughes (1958); Becker (1952, 1982); Strauss (1959; Glaser and Strauss 1965, 1971; Strauss et al. 1964); and Goffman (1959, 19 61a, 1961b, 1967, 1974).

Roles, Role Relationships, and Encounters

Drawing on the work of the anthropologist, Siegfried Nadel (1957), I find it useful to view roles as having two analytically separate, but empirically entwined components.[3] Nadel proposed that there are

---

[2] When studying technologies, I have always been attuned first to behaviors and second to meanings. This stems, in part, from the fact that I was exposed to the work of Roger Barker (1963) before I was exposed to the interpretive stance epitomized in my mind by Geertz (1973). I subscribe to the notion that we are what we do and am skeptical of the proposition that we are what we think. Of course, what we think my well influence what we do; therefore, what we think is certainly important, but doing usually has more concrete social ramifications than thinking.

[3] My discussion of how we can make use of Nadel's ideas draws heavily from (Barley 1990)

two types of roles: relational and non-relational. He argued that people cannot play relational roles without a specific alter. Hence, there can be no son without a mother, no debtor without a creditor, and no subordinate without a superior. In contrast, non-relational roles require no specific partners. According to Nadel, an actor in a non-relational role need only engage in that bundle of behaviors deemed by members of a culture to be characteristic of the role. Hence, there are no particular alters for the butcher, the baker, or the candlestick maker. To perform these roles it is sufficient that the first butchers, the second bakes, and the third makes candles.

Unlike Nadel, I do not believe that roles can be typed so neatly. In fact, I submit that his concepts are not particularly useful if we treat them as a typology of roles. For example, the butcher's role clearly depends on how he or she treats animals for money, but if the butcher had no customers his or her activities would merit another name as well as another place in society, and, most likely, the services of a lawyer and a psychiatrist. Similarly, mothers not only stand in a unique relation to their sons and daughters; they also have a unique relationship, however minimal or distant, to a father. Moreover, mothers must also perform a minimal set of culturally expected duties if they are to fulfill their role in any but the biological sense of the term. Thus, rather than adopt Nadel's typology of roles, I contend that it is more useful to think of all roles as bundles of non-relational and relational elements that can be separated analytically when it is useful to do so. From this perspective the non-relational aspects of a role can be viewed as the set of recurrent activities that fall within the purview of a person who assumes the role—for our purposes, the practices associated with assuming a position in a division of labor. Because the non-relational elements of a role include tasks and skills (what one does and how one does it), technologies are likely to shape most directly the non-relational aspects of a work role.

Yet few of us perform tasks that are totally independent of other people. In fact, many jobs consist primarily of interacting with people. Even in work structures that Thompson (1967) called "pooled interdependence" (such as a secretarial pool or a call center), people have co-workers as well as supervisors. Hence, what one does and how one does it will influence with whom one interacts as well as how those interactions unfold.[4] For this reason, technologically induced changes in the non-relational aspects of a work role may spill over into the role's relational elements. For example, altering a role's tasks or techniques may narrow or expand the range of a person's role set, shift the nature of his or her dependencies, or affect the frequency and content of his or her interactions. The sneaky qualification here is "may."

It is entirely possible for a new technology to change the non-relational aspects of a work role dramatically, without significantly altering the role's relational elements. Consider, for example, the work of professors and administrative assistants (once called secretaries). As recently as the 1980s administrative assistants answered phones, interacted with students, kept records of accounts, filed documents, and typed letters, memos, and manuscripts for faculty (who often wrote the first drafts by hand). Today, administrative assistants continue to interact with students, and answer phones (usually not for professors), but they almost never type documents for professors. Professors now use word processing programs to create and revise their own documents. As in the past, administrative assistants continue to manage accounts, but they do so using computerized spreadsheets and web-based forms. Because of the increased efficiency of producing and storing documents and because faculty have assumed the tasks of producing their own documents and answering their own phones, universities now employ fewer administrative assistants, and some of those who remain have acquired new skills and tasks, such as maintaining websites and building spreadsheets. Yet administrative assistants and faculty continue to have roughly the same relationship as they had in the past. There is no doubt about who has greater status or who works for whom. In short, what administrative assistants and professors do and how they do it (the non-relational parts of their role) have changed considerably because of technological change, but the relational aspects of their roles remain the same, although professors and administrative assistants may interact less frequently.

---

[4] I shall discuss the unfolding later in this section under what I call, following Goffman, the "encounter."

Role relationships could be defined structurally as a network analyst or a job description might, for example, with phrases such as "X reports to Y," "A seeks advice from B," or "Project managers coordinate the work of the software developers, electrical engineers, mechanical engineers, and marketers assigned to a project." Such statements tell us something about who interacts with whom. However, from such statements we do not learn much about what actually happens when "X reports," "A seeks advice," or "project managers coordinate." This is because role relations imply lines of situated, yet structured actions and interactions that unfold in time as two or more people play out their respective roles. To understand how technologies change organizations by changing roles, we would be wise to attend to how roles are played. Following Goffman (1961b), let us call an instance of role playing between two or more people an "encounter." In encounters, all actors enact some role following at least a minimal set of *cultural expectations about how, when, where, with what, and with whom the role should be played.* In short, to understand how technology becomes entwined with and alters ways of organizing, we need to attend to the *situated, patterned, and recurrent ways of behaving and interacting* that mark a particular context: what Irving Goffman (1983) called an interaction order.[5]

Over his career, Goffman sought to understand why social life is so well ordered and to reveal why encounters of particular types (a lecture, a game of checkers, and so on) unfold so similarly regardless of who is involved.[6] The key, for Goffman, did not involve understanding individual-level sense-making such as occurs when people attempt to comprehend the unexpected, but rather sense-making and behavior that are contingent on a "definition of the situation" or what he later called a "frame" (Goffman, 1959, 1974).[7]

---

[5] The discussion of encounters in the following eight paragraphs draws heavily on Barley (2015: 9–10)

[6] Although some scholars claim that Goffman was primarily interested in the construction of selves, identities, and the negotiation and maintenance of face, his agenda was ultimately to explain the social structure of everyday life (Collins 1988, 2004; Giddens and Turner 1987). It is on Goffman's structural agenda that I draw in this chapter.

[7] One should not confuse Goffman's use of "frame" with its use by contemporary institutional theorists who often use the term as a synonym for a logic. Among institutionalists, frame has the connotation of a rhetoric, concept, or ideology. In Goffman's terminology,

Once we have defined or labeled the situation, we know how to act and play our role in the situation.

Although definitions of a situation are sometimes negotiated on the spot (Weick 1993), as when we explicitly ask each other, "What's going on here?," they are not usually constructed *de novo* and they are not unconstrained. Constraints arise from how activity is embedded in and arises from a context. Once we classify some goings-on, we know how to act because we treat the goings-on as a case of some typical social situation. Goffman argued that frames are layered contexts that configure the lines of action that participants can take in a particular type of encounter. Importantly, these layers include the "physical framework," by which Goffman meant artifacts and natural phenomena, as well as the "social framework," by which he meant the other actors who are present as well as the institutional setting (such as a theater, a church, a lecture hall) that provides specific rules for acting (Goffman, 1974: 21–5). People cannot ignore the physical framework any more than they can ignore the social. Once a situation is defined, people engage in "guided doings" (Goffman, 1974: 23): They construct a line of action within the confines of the possibilities set by physical and social frameworks (Pinch, 2010; Darr and Pinch, 2013).

Like other dramaturgical sociologists (Messinger, Sampson, and Towne, 1962; Turner and Edgley, 1976; Wilshire, 1982), Goffman argued that once a situation is defined, encounters can be usefully viewed as theatrical performances.[8] The dramaturgical eye draws the analyst's attention to specific elements of the encounter and its setting which support and structure the unfolding line of action. Figure 2.2 depicts the key elements of an encounter.

Scripts are the pivotal element of any dramaturgical account. A script organizes and typifies action and interaction by defining how

---

frames define classes of situations which exist at a much more contextual level of analysis. Goffman used frame in the sense of a picture frame, a set of cues or parameters that set the situation at hand off from its surroundings and that define acceptable lines of action within the context.

[8] To argue that encounters are rituals or theatrical performances is not to say that actors are somehow disingenuous or that their actions and interactions are contrived, although the scripted aspects of encounters do make possible the kind of dissembling that drew Goffman's attention to confidence artists.

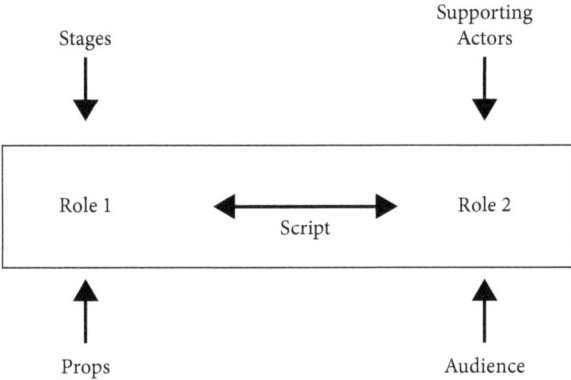

Figure 2.2. The Dramaturgical Elements of an Encounter

parties to an encounter should play their roles. The vast majority of encounters have scripts even if they are as minimal as the two-turn exchange of greetings between strangers acknowledging each other's presence on a sidewalk (Sacks, Schegloff, and Jefferson, 1974; Schegloff, 1987). One can think of a script as the plot of a recurrent activity that defines the essentials of the parts that participants must play. People can certainly improvise around a script giving a particular encounter a sense of uniqueness. But even if we acknowledge improvising, the flow of the script must be generally sustained if the encounter is to conform to the definition of the situation. Consider going to a restaurant (Schank and Abelson, 1977). When we go to a restaurant, the stream of action unfolds in a relatively predictable order: the maître d' greets us; we ask for a table; the maître d' seats us; the waiter greets us at the table and hands us a menu; the waiter asks if we'd like something to drink; and so on through dessert and paying the bill. As Garfinkel (1967) demonstrated, when scripts are breached—as would occur if the waiter immediately asked if we would like dessert—we become disoriented, even anxious or angry.

Scripts are both cognitive and behavioral phenomena. Cognitively, scripts are our expectations about how things ought to go, given a definition of the situation. When our expectations are fulfilled, we have the sense that "Nothing out of the ordinary is going on here," although we are not likely to be aware that we are experiencing a sense of the usual. Behaviorally, as the dramaturgists emphasize, scripts are culturally but loosely prescribed sequences of behaviors and interactions associated with a type of encounter. People enact or play their respective roles by making moves at appropriate points in an interaction's unfolding (see Pentland, 1992). A scripts' enactment is often supported by, and may require, a specific stage (a restaurant in the evening), a set of props (tables, plates, candles, and a corkscrew), supporting actors (bartenders, busboys), and sometimes an audience (other diners). When an encounter's dramaturgical elements change, scripts may change, and when scripts change, by definition, the interaction order has changed. For example, asking customers to pay before they eat food that was cooked before they ordered it and then asking them to seat themselves and bus their own tables turns a fine restaurant into a fast-food establishment.

Scripts unfold through *moves*, the actions that a second party takes in response to a first party's actions that, in turn, serve as clues and cues for the next action the first party should take (see Collins, 2004). Moves are strategic in intent; through them actors hope to achieve objectives. Encounters also vary by their *footing*, the stance that actors assume toward each other as they make their moves. Footing sets the tone of the encounter. When you hand an immigration official your passport, you engage in roughly the same behavior as when you purchase a movie ticket: in both cases you hand a legal document to someone in a glass booth who then offers admission beyond. In the case of immigration, however, you present yourself as a *supplicant* for admission, which the official can question and deny. In the case of going to a movie, you present yourself as a *customer*, and you do not expect to be denied admission unless the show is sold out, your cash is counterfeit, or your credit card is deemed invalid by a computerized validation system. To appreciate the difference in footing one need only consider the relative consequences of expressing irritation at the ticket seller versus an immigration official. In these instances, footing pertains to who has more power in the dyad.

As mentioned above, other dramaturgical elements of an encounter support and shape the scripted line of action and the moves by which actors play their parts. Like plays, encounters occur on some *stage* and sometimes in front of some *audience*, which might be present or imagined. The stage is bounded in time and space, demarking the region within which an encounter of a particular type can and should occur. Some settings have front stages which are available to the audience and backstages which are not. Moreover, stages usually contain *props* that buttress the line of action and the actors' performance of their roles. Stages and props are the material artifacts (some of which we call technologies) that actors employ as they play out encounters. One of Goffman's key points is that actors cannot sustain an encounter of a given type without the right stage and the right props. If you change the props or the stage, you risk changing the definition of the situation. For instance, dining in a restaurant requires the presence of tables, chairs, plates, forks, knives, and spoons. In restaurants people do not stand or eat with their hands. At picnics they may. The point is that technologies and other artifacts help structure our definition of the situation, the scripts that guide the encounter, and the way we play our roles.

In sum, a dramaturgical analysis of technological change differs from other approaches to studying technology. Rather than make the technology the center of attention, it shifts attention to the system of actors, actions, and interactions in which the technology is embedded. We no longer are simply interested in how a technology changes the way we do some task, a pattern of communication, how people conceptualize the technology, or even the practices that emerge around a technology's use. Instead, dramaturgical analysis attends to the entire milieu that flows from and sustains a definition of the situation and plausible lines of action and interaction. The technology is entwined with the line of action. If anything, it is the script rather than the technology that lies at the core of the analysis. *The analyst's job is to determine whether the presence of a technology has somehow reconfigured the scripts, the stage, the props, the moves that actors make, or the encounter's footing in ways that sustain an altered or different line of action.*

Because scripts, by definition, encode role relations, dramaturgical analysis brings the playing of roles and, hence, role relations to center stage. To have social consequences for organizing, a technology must reconfigure some combination of the scripts, moves, actors, footing, props, and stages that structure an interaction ritual to transform the way a type of encounter unfolds. Only by doing so will a technology shift the relational aspects of a role, which are precursors to changes in the networks that structure the organization.

To illustrate how technologically occasioned change reverberates from changes in the non-relational aspects of a role to the affect encounters and role relations and then on to social networks, I will first revisit my early work on radiological imaging technologies (Barley 1986, 1990). To keep the discussion brief, I will confine my description to radiological technologists who produce standard X-rays and assist in fluoroscopy and to sonographers. As a second illustration, I will draw on a more recent study of how the Internet changes relations between car salesmen and their customers.

Computerized Imaging

During the late 1970s and early 1980s computerized medical imaging modalities began to diffuse from research labs and teaching hospitals into community hospitals in the United States.[9] Previously, radiology departments specialized exclusively in the production of radiographs (X-rays) and fluoroscopic exams. Radiographs and fluoroscopy relied on cathode ray tubes, which had been used for medical imaging since before World War I. The difference between the two was that radiographs were static images taken at a single point in time, while fluoroscopy allowed continuous imaging and was used primarily to track the movement of barium (a radiopaque substance) through either the upper or lower gastrointestinal tract. Importantly, neither technology involved digitized data or computers to construct images.[10] They were analog devices that used photographic film to record images in much the same way that cameras recorded images prior to digital photography.

---

[9] In radiology a type of imaging technology is called a modality. I will use the emic term when appropriate.

[10] This was true in the early 1980s when I did my research. Today radiographs and fluoroscopy have both been digitized.

In 1982, when I set out to study computerized imaging, three digital modalities were being used in community hospitals: ultrasound, computerized tomography (CT scanning), and digital subtraction angiography (DSA).[11] I designed my study to compare and contrast the use of sonography, DSA, and CT scanners with standard X-rays and fluoroscopy in two community hospitals. I picked the hospitals because they were in the process of adopting their first whole-body CT scanner, but my primary interest was in documenting differences in the social organization and use of computerized versus traditional imaging technologies. Within the radiology departments that I studied there were seven primary roles: radiologists who were MDs; radiological technologists who held associate's or bachelor's degrees; departmental administrators who were former technologists; nurses who assisted radiologists in angiography; clerks who scheduled patients, transcribed the radiologist's readings, and booked appointments; orderlies who transported patients to and from the radiology departments; and darkroom technicians who developed the films on which images were recorded. The technologists were further distinguished by the technology they operated, with four distinct positions or specialties: sonographers, CT techs, special procedures techs (angiography), and X-ray techs who produced radiographs and assisted with fluoroscopy. For a year I shadowed radiologists and technologists almost daily in one or the other hospital to ensure that I recorded multiple instances of what a radiologist's or a technologist's work consisted of and how they played their roles when using each type of technology. In Goffman's terminology, examinations were nicely bounded encounters.

The radiology departments' standard job descriptions prescribed role relationships as depicted in the social network in Figure 2.3. The important point to note is that in Figure 2.3 there are no substantive differences in the positions of the various technologists: In the graph all technologists are structurally equivalent (White,

---

[11] Magnetic Resonance Imaging (then called Nuclear Magnetic Resonance—NMR) and Positron Emission Tomography (PET) were emerging but were located only in research labs and major medical centers. NMR eventually became known as MRI because the term "nuclear" unnerved patients. MRI (like ultrasound) employs no radioactive materials (it makes use of magnetic fields), whereas patients are exposed to radioactivity during radiography, fluoroscopy, computerized tomography, and angiography.

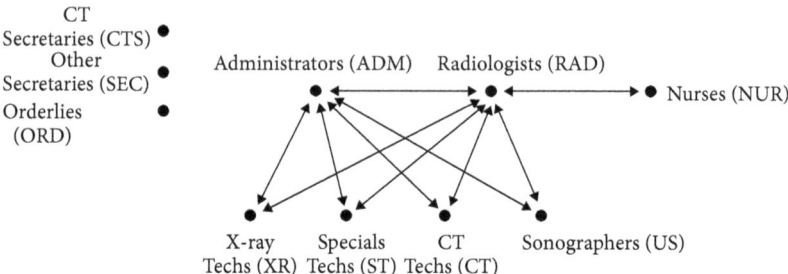

Figure 2.3. Role Structure of a Radiology Department According to Job Descriptions

Boorman, and Breiger, 1976; Burt, 1982). All radiologists were also formally identical. As organizational theorists might expect, Figure 2.3 is an inaccurate image of role relationships between radiologists and technologists in the departments. Consequently, as we shall see, the graph in Figure 2.3 is an inaccurate depiction of the departments' structures. The discrepancies can be traced to differences in imaging modalities, in particular, whether the modality was or was not computerized and whether the production of images entailed a cybernetic process. These differences affected the non-relational and relational aspects of the roles of both radiologists and technologists. As a result, the structure and tenor of the encounters between technologists and radiologists varied across modalities.

I shall outline the scripts that radiologists and technologists followed when producing "routines" (what we would call X -rays) and sonograms. I explicate these two because they offer the sharpest contrast between the non-relational and relational roles of members of an occupation that used an analog or a computerized modality respectively. The roles played by technologists doing other examinations with analog technologies were similar to those doing "routines," and the techs operating the other computerized modalities were similar to sonographers along the dimensions I discuss below, but their differences from X -ray techs were less extreme than those of sonographers.[12]

---

[12] See the tables in Barley (1990) for quantitative evidence for these claims.

I shall also confine my comments to encounters between radiologists and technologists, while ignoring encounters between radiologists, between technologists, and between members of both occupations and patients. I do so because the changes occasioned by computerized modalities arose primarily from shifts in relations between radiologists and technologists. Nevertheless, one should note that the exact nature of the roles that technologists and radiologists played and the relations between them varied not only by modality but by the type of study being done with the modality.[13]

Before I turn to roles and encounters, several facts about the distribution of expertise in the two departments are important to mention. When the departments adopted each computerized modality, the radiologists were aware that they could not competently interpret the images it produced. Thus, the departments hired radiologists who were experienced in reading the new modality's images. Because the departments purchased the modalities at different times, there arose differences in which radiologists could do what. All could read X -rays and fluoroscopic images. Fewer could interpret sonograms and images produced during special procedures, and even fewer could read CT scans. As I have explained elsewhere (Barley, 1990), the radiologists nevertheless rotated through each technology with the exception of special procedures, which required radiologists to possess surgical as well as interpretive skills. The radiologists knew when they could not render a competent reading and, therefore, when they could not provide an interpretation to a referring physician until it was checked by a radiologist who was proficient in the modality.[14] Doing otherwise would endanger a patient and invite malpractice suits. Accordingly, the younger radiologists carried a heavier workload; they were responsible for reading images for the modality to which they were assigned on that day as well as the images produced on other modalities when those modalities were staffed by an inexperienced radiologist.

[13] Readers interested in this variation should consult Barley (1984), which lays out the work roles associated with modalities and types of exams in excruciating detail.

[14] A referring physician was the doctor who requested radiological examinations to gather further data on patients and their conditions.

Similarly, when adopting computerized modalities, the departments had either to hire technologists who were already proficient at operating the modality or to train one or more X -ray techs in the new technology. Each of the departments took both paths: They hired experienced technologists and trained a few former X -ray techs under the tutelage of the more experienced technologists. Thus, for each new modality the radiology departments created a new occupation or position for technologists operating the new modality. They did not, however, create distinct positions for the radiologists.

Scripts for Doing Routines and Sonograms

Routines: The micro-details of producing a routine X-ray varied according to the part of the body being imaged (leg, knee, chest, abdomen, etc.). Nevertheless, all routines unfolded in roughly the same sequence, and every instance of a type of exam followed the same script as depicted in Figure 2.4. The examination (and the script) began when the technician retrieved a requisition and a "darkroom card" from the clerk who staffed the department's reception desk. The card would be used to label the patient's films (in the figure, "P" stands for patient). The requisition included information on the part of the body to be imaged, the number of films to be taken, and sometimes the reason for the study. The tech greeted outpatients in the waiting room and inpatients at the wheelchair or gurney in or on which orderlies had brought the patient to the department. If an outpatient was female, the tech would lead her to a dressing room and ask her to change into a hospital gown or "johnny." Techs only asked male outpatients to don a johnny if they would be imaging the lower abdomen or legs. Inpatients never needed to change garb because they always wore hospital gowns.

Once patients were in a state of acceptable dress, the tech led or wheeled them into the examination room and shut the door after entering. With the exception of chest films, which were done with the patient standing or films of the wrist and hand which were done with patients sitting in a chair resting their hands on the X -ray

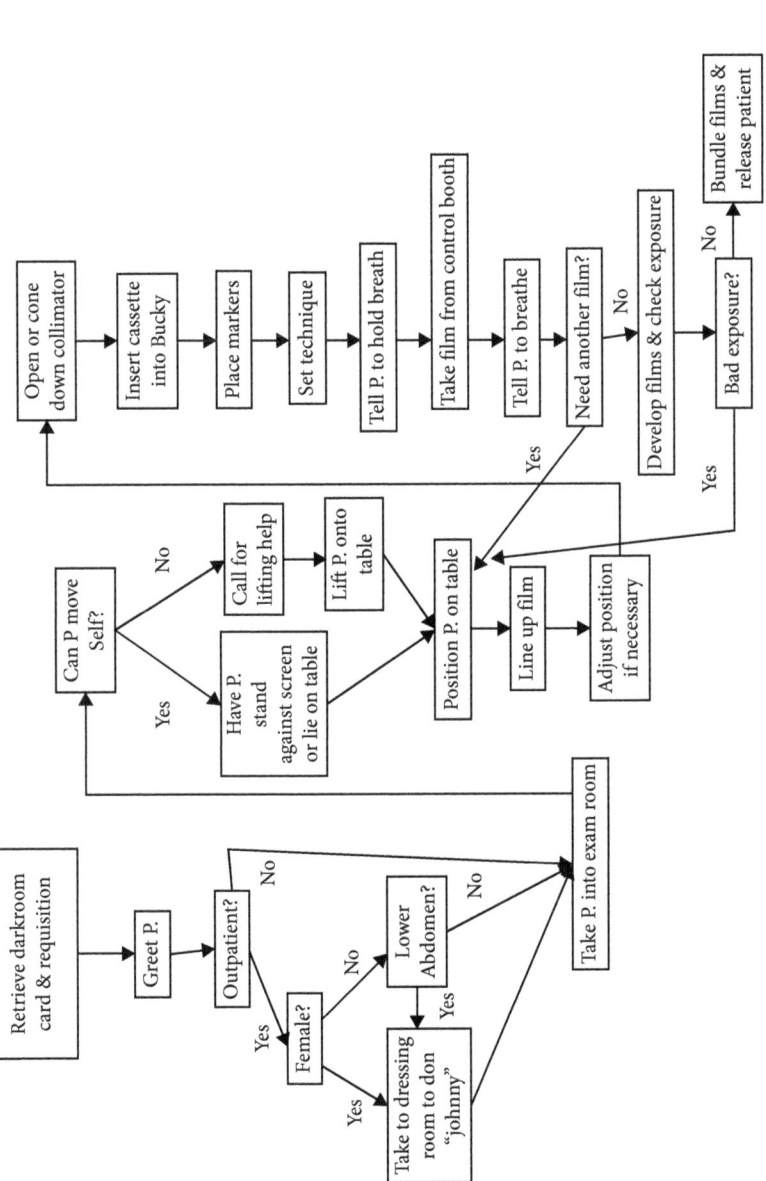

Figure 2.4. Script for Doing a Routine X-Ray

equipment's table, all studies were done with the patient lying on the table. The next order of business was to move the patient to the place where films were to be taken. In the case of a chest film, patients were told to stand next to a screen. Otherwise, techs asked patients to lie on the table, which posed no problem when the patient was a "walkie-talkie," that is, ambulatory.[15] Patients in wheelchairs or on gurneys posed greater difficulties, especially if they were infirm or incognizant. If a patient in a wheelchair was a surgical patient, seriously ill, or elderly, the tech assisted, coached, and encouraged them out of the chair and onto the table. If the patient was on a gurney, the tech moved the gurney next to the examination table and adjusted the latter's height so the patient could slide from the gurney onto the table. If the patient could not wiggle onto the table, the tech summoned assistance by walking into the hallway and calling for "lifting help." On hearing the call, three of more orderlies and technologists entered the room and took up positions around the table and gurney. On the tech's command, the group literally lifted the patient onto the table using the gurney's sheet as a litter. Once on the table, the help quickly left.

With the patient on the table (or standing against the screen), the technologist began positioning the patient, the first of two technical skills on which the occupation of radiological technologist rested. Positioning involved orienting the patient's body in relation to the X-ray tube so that the "area of interest" could be filmed from a prescribed point of view. When positioning the patient, the technologist molded the patient's body into a pose either by coaxing or by manually manipulating the patient's body. With the patient posed, the tech refined the position by "lining up" the film using the X-ray tube's "field lamps," which projected a rectangular area of illumination bisected by horizontal and vertical shadows cast by lead strips in a pattern resembling the crosshairs of a rifle scope. The tech adjusted the patient's position until the shadows fell on anatomical features that served as the position's landmarks. If the table had a "floating top," the tech could adjust the positioning by

---

[15]   In the course of the descriptions of scripts, terms in quotes are emic expressions used by my informants.

stepping on a foot pedal and moving the table top. Otherwise, the tech had to adjust the position manually by manipulating the patient's body.

With the patient positioned, the tech then "opened" or "coned down" the field of view by shifting the tube's "collimators," lead strips located in the tube's housing. Collimators blocked electrons and shaped the size of the field to the film. The size of the film was determined by the area of the body to be imaged. Most X-ray tables used films housed in "cassettes"—thin rectangular metal boxes slightly larger than the film itself. Before an X-ray could be taken, the tech had to insert the cassette into a holder beneath the table known as a "bucky" or into a similar holder behind a standing screen. With the cassette in the bucky, the tech taped a marker containing lead letters signifying the tech's initials to the cassette's lower corner to identify who took the film. The markers also contained a lead "R" or "L" indicating the right or left of the patient's body.

While positioning, the tech talked more or less constantly with the patient about the pose the patient should assume. When satisfied, the tech told the patient they were ready to take the film and that they should not move until the film was over. The tech now walked to the tube's control panel and "set a technique," which entailed selecting the kilovolts and milliamperes that defined the electrical current applied to the X-ray tube's cathode as well as the number of seconds the tube would fire. Setting a technique was the second skill on which the radiological technologist's occupation rested. Loosely speaking, a technique determined how long the tube emitted how many electrons traveling at what speeds. If the tech did not set the technique correctly, the film would be over- or underexposed.

After setting the technique, the tech entered a lead-lined control booth, told the patient to hold their breadth, and took the film. After taking the film, the tech left the control booth, told the patient to breathe, and removed the cassette from the bucky. The tech repeated the process from positioning onward until they had taken the requisite number of films.

Next the tech took the cassettes to the darkroom's passboxes (two-way cabinets in a wall) in a hallway from which the darkroom technician retrieved the cassettes, opened them, and put the film in a film processor. Like passboxes, the film processors were also imbedded in the darkroom's wall. A system of rollers drew the film through photographic chemicals and a drying process. Once developed, the film fell into a bin in an adjacent room. The tech removed the films from the bin, hung them on light boxes, checked for adequate exposure, and ensured that the area of interest was properly centered using anatomical landmarks. If the films were adequate, the tech placed them in a folder for a clerk to take to a radiologist, who would read them later. The tech then dismissed the patient, which entailed moving the patient back onto a gurney, if necessary, in reverse order to what had been done at the beginning of the exam. Notice that neither the technologist nor the patient interacted with the radiologist and that the technologist's inter-action with the patient was minimal and confined largely to greetings, giving directions on how to position their body, and sometimes small talk. In short, the techs primarily gave patients orders.[16]

When reading routines, radiologists stationed themselves in a reading room equipped with light boxes. To read a study, the radi-ologists hung the films on a light box and examined them for evi-dence of pathology. Radiologists might consult a colleague if they were uncertain about or intrigued by what they were seeing, but usually they had no need to consult. After studying the films, the radiologists dictated an interpretation (called a "reading") into a Dictaphone. A transcriptionist eventually retrieved the recording and transcribed the reading onto a paper form. The radiologist then proceeded to the next study. At unpredictable times, referring phys-icians might call or visit the radiologist to see if a study had been completed or to discuss the results of a study. It took radiologists between one and five minutes to read a routine study, with the dis-tribution skewed toward lower time intervals.

Sonograms. Sonograms also followed a script (see Figure 2.5), but the script admitted more variations depending on the patient

---

[16] During an intravenous pyelogram or fluoroscopy technologists did interact with radi-ologists, but these interactions were highly scripted. For a detailed account of how these other procedures were scripted, see Barley (1984).

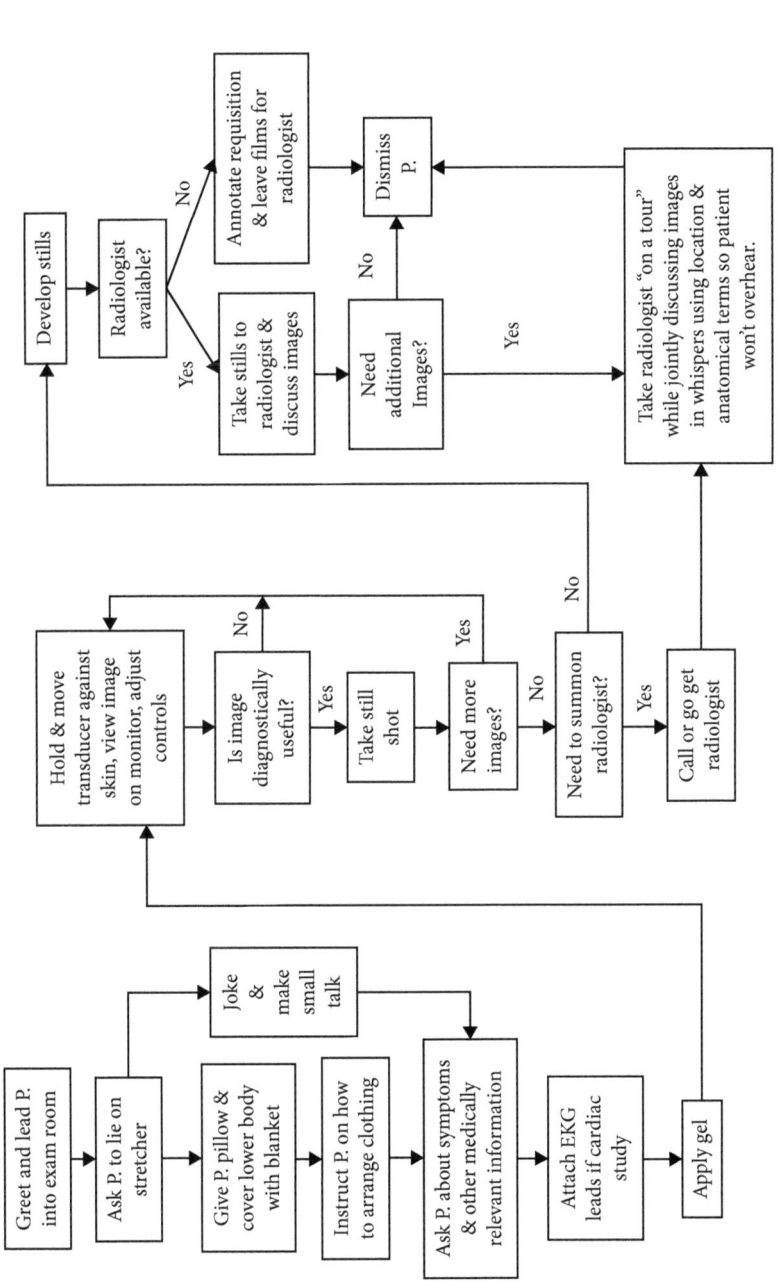

Figure 2.5. Script for Doing a Sonogram

and the difficulty the sonographer had in capturing the images to produce a diagnostically viable study.[17] Sonography was used to diagnose a number of conditions ranging from pregnancy to aneurisms. Sonographers produced sonograms by holding a hand-held transducer against the patient's skin. During the course of a study the sonographer moved the transducer while watching moving images on a video monitor. The equipment that I observed allowed the sonographer to record the images on a videocassette to capture the exam in motion as well as on photographic film enclosed in a cassette similar to those used in X-ray to create "still" images.

At the start of a sonogram, the sonographer greeted the patient, introduced him/herself, and led the patient into the ultrasound room and asked the patient to lie on a stretcher. Once the patient mounted the table, the sonographer offered a pillow and covered the patient's lower body with a blanket, less to guard the patient from the cold than to guard the patient's "decency." With the blanket in place, the sonographer instructed the patients to lower their pants or skirt or to raise their shirt, blouse, or dress.[18] While patients arranged themselves on the stretcher, sonographers always struck up conversations marked by the exchange of pleasantries, small jokes appropriate to ultrasound ("Did they give you enough to drink" (to fill the bladder);[19] "So we're going to see how your ticker tocks") and other forms of small talk. Inevitably, however, sonographers gradually led patients into extensive discussions of their symptoms. They queried patients for specific locations of pain, previous diagnoses provided by physicians, and an experiential account of the condition that brought the patient to the study. Sonographers were adamant that without assessing the patient's symptoms they could not conduct a study competently, because

---

[17] Readers familiar with ultrasound will note that I am describing an M-mode (motion) rather than a B-mode (brightness) study. At the time of my research, the equipment at both hospitals allowed only B-modes and M-modes. I have chosen to describe M-mode because they were more common and because they emphasize the real-time nature of the technology.

[18] If the patient was female and having a cardiac study, she would have previously been asked to change into a johnny, and the sonographer would discreetly open the midline of the gown to reveal her sternum without exposing her breasts.

[19] Sound waves travel through water, allowing the sonographer to better image other organs such as the uterus. Water-filled structures appear black on an ultrasound image.

symptoms were clues to the patient's potential problem and consequently, to which organs and what type of lesions the sonographer should be alert to. Before beginning to collect data, the sonographer covered the area of the body to be imaged with a gel to create a liquid interface between the head of the transducer which improved the transfer of sound from the transducer to the patient's body. If the study was a cardiac exam, the sonographer attached EKG pads to the patient's chest and linked the pads to an EKG monitor before proceeding.

A bank of real-time ultrasound equipment stood to the side of the examination table. The compact ensemble contained a computer, a keyboard, a video monitor, a camera, and a videotape recorder, all enclosed in a metal cabinet that stood approximately 5 feet high and 3.5 feet wide. After applying the gel, the sonographer turned to the keyboard and typed in the patient's identifying information as well as the perspective from while the first images were taken. The sonographer then dimmed the lighting in the room to more easily view the monitor. The sonographer stood next to the patient and held the transducer against the patient's body. Because sonographers had to watch the monitor to conduct the study, they faced the bank of equipment and used their free hand to adjust the parameters of the equipment (for example, "gain"[20]) and to change labels on the monitor. The exam progressed as the sonographer shifted from one perspective to another and from one organ to another by moving the transducer along the patient's body or rotating it against the skin. What sonographers saw or wished to see on the monitor dictated how they moved the transducer from one instant to the next. From time to time, the sonographer would freeze the video image and take a still shot of the screen. Although there were protocols that stated which organs and which

---

[20] Most people think of gain as a brightness adjuster, and while it's true that turning your gain up will brighten the image, it's helpful to understand how it actually works. Gain is a *uniform amplification of the ultrasonic signal* that is returning to the transducer after it travels through the tissue. So rather than brightening the monitor, the image on the screen is whitened by a uniform margin, as though the returning signal is stronger than it is, to make it easier to see.

(https://www.eimedical.com/blog/understanding-gain-in-ultrasound, accessed December 29, 2019)

perspectives a sonographer had to film during any procedure, the number of images was left to the sonographer's discretion and varied with the pathology that the sonographer discovered. Sonographers claimed that it was their responsibility to capture enough data so that the films convincingly showed the radiologist that a particular condition did or did not exist. Consequently, sonographers attempted to triangulate onto areas of interest from multiple perspectives.

When the sonographer finished filming, the exam proceeded in one of two ways. First, the sonographer might summon the radiologist on ultrasound duty to the examination room by a phone call or over the intercom. In one hospital I studied, summoning was routine. Once the radiologist arrived, he positioned himself beside the sonographer in front of the monitor. In most cases the sonographer now provided the radiologist with a real-time "tour of the anatomy and pathology." The sonographer did so by verbally noting the orientation of the transducer and by naming the particular anatomical structures displaying on the monitor. The radiologist often directed the sonographer to image certain organs or provide views from certain perspectives. Radiologists who were more competent at ultrasound occasionally manipulated the transducer themselves. Conversations between the sonographer and radiologist as they jointly viewed the images transpired in whispers. Often, the two held short debates about what they saw on the screen. They were as likely to discuss pathology as anatomy.

In the other hospital I studied, the radiologists almost never came to the ultrasound room. Rather, when the filming was complete, the sonographers took leave of the patients by saying that they were going to develop the films (the stills) and that they would return in a few minutes. After developing the films (as described above for X-ray techs), the sonographers might or might not take the films to a radiologist to view. The decision depended largely on which radiologist was on duty and which sonographer was performing the exam. If the sonographer decided to have a radiologist view the films, he or she carried the films to the radiologist's office and hung them on the radiologist's light box. After hanging the films, the sonographer gave the radiologist only enough information to orient him to the films and the patient's condition. The radiologist then read the films and as he read, he asked the sonographer questions.[21] If the sonographer agreed with the radiologist's interpretation, he or she said "Uh-huh" or "That's what I thought too." If the sonographer disagreed, he or she called the radiologist's attention to other data that he or she thought incompatible with the radiologist's interpretation. In the same manner, the radiologist challenged the sonographer's interpretations. If the two could not reach agreement, they might return to the ultrasound room and redo the exam jointly. When sonographers did not immediately take films to the radiologist, they noted on the requisition what they thought the films revealed.

Once the radiologist had either conducted a study or the sonographer had delivered the films to the radiologist, the sonographer returned to the examination room and told the patient the exam was over. The sonographer then wiped the gel from the patient's body with "chucks," an absorbent, disposable napkin. Usually the sonographer told the patient that the gel did not stain clothing. With the gel removed, the patient dressed and left the examination room.

Differences in Roles and Role Relations

The contrast between the scripts for routine X-rays and sonograms points to key differences in the social organization of encounters in traditional and computerized studies. In routine X-rays radiologists and technologists rarely interacted, whereas interactions between sonographers and radiologists were common. Second, and more importantly, radiologists almost never told an X-ray technician what they saw in films. In contrast, radiologists and sonographers carried on detailed conversations about the anatomy and, more importantly, the pathology revealed by a study. This difference is critical, because the American Radiological Association banned technologists from being taught how to interpret films (Barley, 1986); yet the ability to interpret was central to the

---

[21] In both departments, all radiologists were male.

sonographer's ability to produce a medically useful study. Sonographers needed to be able to read images, because, unlike routines and other studies done by X-ray technologists, a sonographer's next move was dictated by what he or she just saw on the video monitor. In short, ultrasound was a cybernetic technology: The images served as feedback as well as stimuli for the sonographer's next action.

Aside from knowing whether a film was under- or overexposed and whether the anatomical area of interest was imaged, films provided X-ray techs with no information on what to do next. For this reason they could produce medically useful studies without needing to understand most of the information they contained. To one degree or another, CT scanning and angiography were also cybernetic: Each image contained information about what should be done next. Hence, CT techs and the techs who did special procedures also had close relations with radiologists, and their conversations also covered discussions of pathology. As I have discussed more extensively elsewhere (Barley, 1990), relations between technologists and radiologists in the computerized technologies were more collegial than in X-ray, where conversations between radiologists and technologists were rarer and consisted of small talk and of the radiologist issuing orders to the technologist, typically using imperative sentence structures. In the computerized modalities, radiologists primarily made statements and asked questions.

To succinctly capture the contrasting role relations of radiologists and technologists in analog and computerized technologies, I asked all members of both departments to complete network questionnaires that asked with whom they conversed about various technical and work-related matters.[22] The resulting graphs are displayed in Figure 2.6. Note first that both of the graphs are quite different from the graph based on job descriptions (Figure 2.3). Second, note that in both hospitals the technologists associated with the computerized modalities are tied to the radiologists and, in only one instance (Urban), to the administrators. In fact, the single tie

---

[22] See Barley (1990) for the precise questions and a more extensive analysis of the network data.

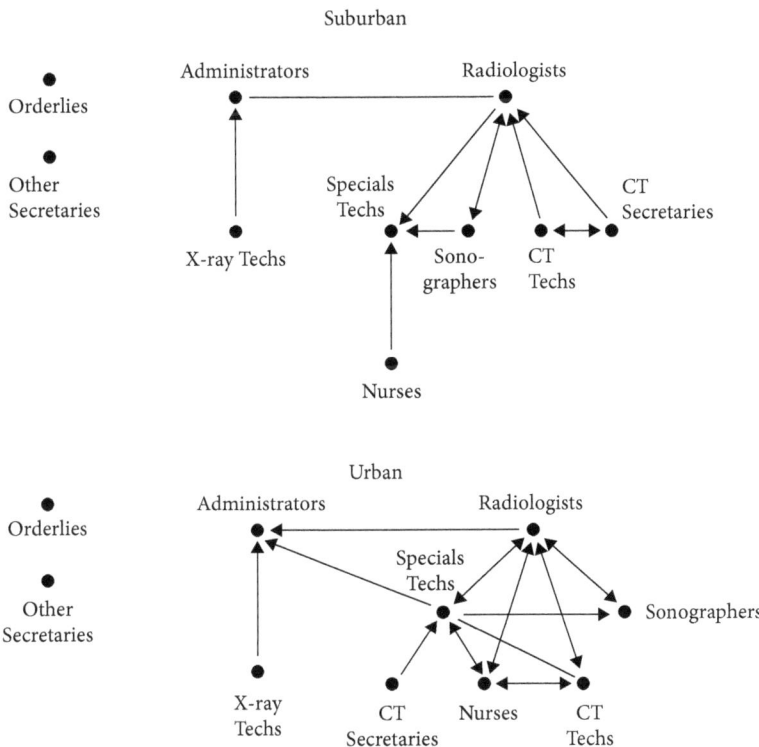

Figure 2.6. Networks of Work Relations at Suburban and Urban Hospitals

between the specials techs and the administrators existed because the administrators at Urban felt they had lost touch with the computerized technologies and, therefore, appointed one of the specials techs to serve as a liaison. Third, note that X-ray techs and radiologists are not directly connected in either department. Finally, and most importantly, the graphs indicate that the two departments had become bifurcated into sub-organizations connected primarily by ties between radiologists and administrators. To all intents and purposes, the administrators ran the X-ray department, while the radiologists ran the computerized technologies. These graphs are evidence that the computerized modalities had altered role relations between technicians, on the one hand, and radiologists and administrators, on the other, thereby changing the department's de facto structure. In neither department was this outcome foreseen, nor did manufacturers tout the outcome as one of the effects that radiology departments would likely experience when they began adopting computerized modalities. It was a second-order effect, as defined in Chapter 1.

Nevertheless, members of the departments realized that two worlds had emerged. At both hospitals, technologists sensed the difference. This was particularly true at Urban. CT techs, specials techs, and sonographers at Urban routinely noted that whereas the X-ray techs were treated like "employees," they were treated as members of a "team." Even the radiologists were aware of the different atmospheres. On the way to lunch one day, two of Urban's radiologists were speaking of the differences between the X-ray techs and the CT techs. One radiologist said to the other, "I see what you mean by the joys of industrial peace." The radiologists concluded that they did not feel nearly as embattled when they were assigned to the newer modalities.

Selling Cars through the Internet

Before the mid-1990s, Americans bought their cars by visiting a dealership franchised to sell a particular manufacturer's vehicles.[23] They examined vehicles either parked on the dealer's lot or displayed in a showroom. They also encountered members of one of the most culturally despicable occupations in the United States: automobile salesmen. In a 2009 Gallup poll, Americans rated only Members of Congress as having lower ethical standards than a car salesman.[24] It is worth remembering that buying a car is the second most expensive purchase that Americans make, and, aside from buying a house (the most expensive purchase), it is the only transaction during which American's expect to bargain. Most American's dislike bargaining, especially with someone they expect to take advantage of them by playing tricks, if not by outright lying.

---

[23] For a more extensive discussion of the data, methods and the scripts, see Barley (2015).
[24] "Honesty and Ethics Poll Find Congress' Image Tarnished" (December 9, 2009), www.gallup.com/poll/124625/honesty-ethics-poll-finds-congress-imagetarnished.aspx, accessed December 29, 2019.

Floor Sales

The buying and selling of a car is a ritualized encounter between customer and salesman.[25] The encounter takes place across a series of stages—the lot, inside a vehicle, and in the showroom—usually in that order. The showroom had a front stage and a backstage. The front stage included the showroom floor and the desks or cubicles assigned to the salesmen. The backstage was where sales managers and the financial people had their offices. Most customers never saw the backstage unless they agreed to purchase a vehicle and eventually talked to the financial people to complete a purchase's paperwork. Over the course of the encounter the salesman employed a number of props to channel the unfolding action. The most important of these were the vehicles themselves, although they also made use of balloons, posters, banners, computers, and even pieces of paper. In the later stage of a sales encounter, the salesman drew on the help of supporting actors—other salesmen and sales managers whose job was to increase the intensity of the encounter by moving from a soft to a hard sell. Customers sometimes also arrived with supporting actors—a friend, a spouse, or another family member–whose job was protect the customer from the salesman's wiles and to express doubt about the wisdom of the purchase and, particularly, about the dealer's asking price.

Most importantly, the encounter unfolded according to a script whose structure varied little by salesman or customer. Figure 2.7 displays the script of a "floor sale," as it is called in the business. The script was constructed from participant observation and tape recordings of sixty-one floor encounters at a Chevrolet and a Toyota dealership in California. There were no significant differences in the scripts at the two dealerships. As Figure 2.7 indicates, the floor script was divided into three acts, what salesman called "landing the customer on a car," "taking a test drive" and "doing the paperwork." The first act usually occurred on the lot, the second always in a vehicle, and the third at the salesman's desk.

---

[25] Although there were a few salespersons in our study who were women, the vast majority were men. To emphasize the gender difference I use the term "salesman" and "salesmen."

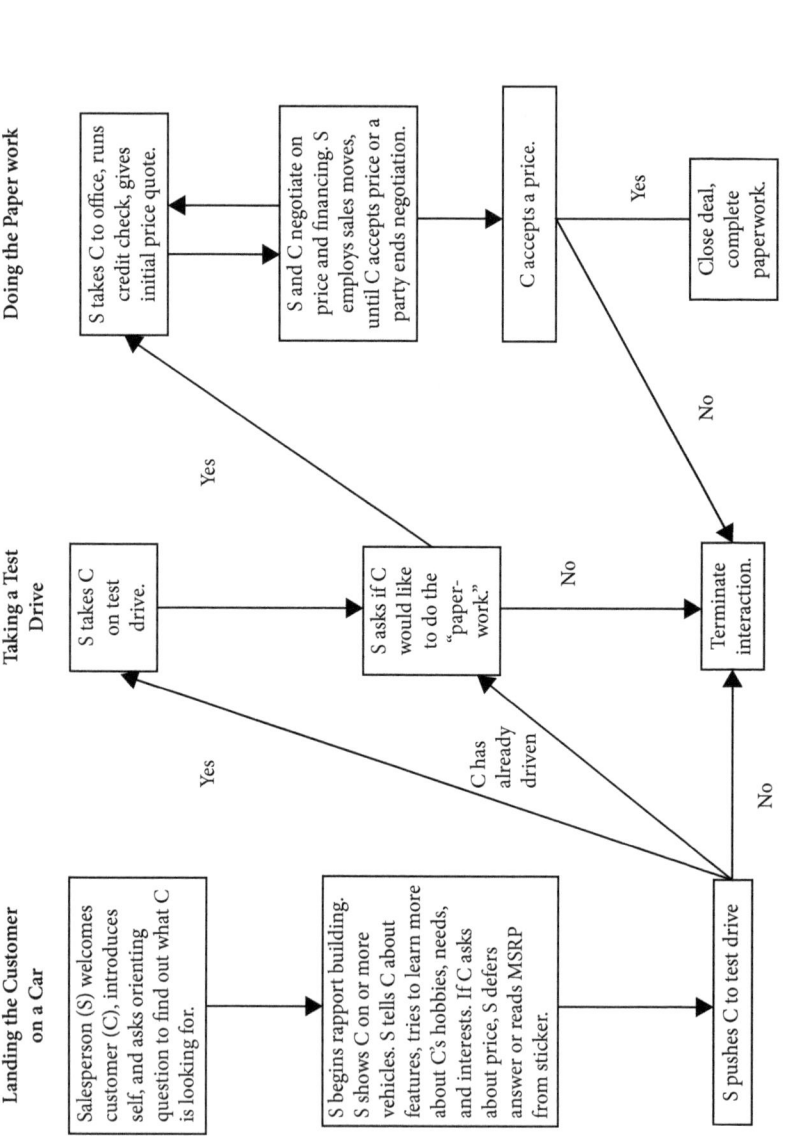

Figure 2.7. The Floor Script

The encounter began when a salesman greeted a customer and asked a question designed to get the customer to state his or her preferences for a model or specific vehicle. Particularly important to the salesman were cues about the customer's readiness to buy and how much he or she could afford to spend. Salesmen attempted to judge financial wherewithal by noting the car in which the customer arrived or by talking with customers about their occupations and employers. Customer's responses could vary from "I'm just looking" to an expression of interest in a particular model in a specific color. It was not uncommon for customers to try to avoid the encounter by saying nothing, by turning their back on and walking away from the salesman, or by speaking a language other than English (although many later spoke English at some point).

If the customer did not end the encounter at this point, the salesman began leading the customer through the lot to a vehicle that came as close to satisfying the customer's stated desires as possible. To "land the customer on a car" (that is, find a car that the customer was at least willing to test drive) the salesman engaged in two entwined lines of action. The first was to "show" the customer specific vehicles and enumerate their features. In the process, the salesman attempted to assess in more detail what the customer wanted from a vehicle, the features that mattered, the potential price range the customer would consider, how the customer would use the vehicle, his or her preferred color, and so on. The salesmen opened doors and trunks and invited the customer to examine the vehicle and its features more closely ("The seats are really comfortable. Why don't you sit behind the wheel?").

While showing vehicles, the salesman attempted to establish "rapport" by learning more about the customer and, when possible, highlighting similarities between the customer and the salesman.[26] For example, if the salesman noted a car seat in the customer's car, he would ask if the customer had a child and then inquire about its age. The salesman would then either profess to have a child of

---

[26] As Cialdini (1984) points out, finding similarities is a basis for establishing liking between two people. It is easier to influence someone if they like you. Interestingly, Cialdini also speaks of car salesmen as individuals who are likely to identify similarities as a basis for influence.

roughly the same age or reminisce about when his children were that age (sometimes even if he was childless). A T-shirt with a firm's logo might lead the salesman to inquire if the customer worked for the firm and to engage in a conversation about the company and what it was like to work there. The logo of a rock band might lead to a conversation about the band's music. If the customer mentioned having a dog, the salesman would steer the conversation toward the dog and then talk about his. Should the customer ask about the price of a vehicle, the salesman would avoid answering or simply note the Manufacturer's Standard Retail Price (MSRP) on the sticker, which both the customer and the salesman knew would not be the selling price.

The objective of the first act of the script was to get the customer to agree to test-drive a vehicle. Accordingly, the first act ended when the salesman asked the customer if he or she would like to "take it for a test drive." At this point many customers refused and ended the encounter. But if the customer agreed, the salesman would photocopy the customer's driver's license, retrieve the vehicle's key, and accompany the customer on a short test drive usually lasting no more than fifteen minutes. During the drive (the second act), the salesman continued to point out features of the car, particularly aspects of the interior, and answer questions that the customer might have. Upon returning to the lot, the salesman would ask the customer if he or she would "like to do the paperwork" (i.e., buy the car). This was the third point at which customers could terminate the encounter and most did. However, if the customer was interested in making the purchase, the encounter entered the script's third act.

It was during the third act that customers and salesmen negotiated the price of the vehicle. At this point the tenor of the interaction between customer and salesman changed, often becoming more adversarial. Salesmen and customers had available to them a number of moves that shaped the flow of the interaction. For instance, salesmen would routinely respond to the customer's opening offer with "I'm not sure my sales manager will let it go for that price." At this point salesmen excused themselves and disappeared backstage, where they reported the customer's offer to the sales manager. Often the salesman dallied backstage, keeping the customer waiting, thereby raising the customer's level of suspense and anxiety. Typically, when salesmen returned, they presented a price higher than the customer offered. Salesmen also attempted to "create urgency" by telling the customer that they could not guarantee the vehicle would not be sold by the time the customer took a day or two to decide or that they could not guarantee the same price on another day. Salesmen frequently told customers that they could not accept an offered price because they needed to "make money on the car." Not infrequently, the salesman "turned the customer," that is, they handed the customer over to another salesman or the sales manager whose job was to create greater pressure on the customer to buy.[27]

Customers also resorted to a set of typical moves during the negotiation. These included threatening to go to another dealer, explicitly disparaging the dealer or the manufacturer, using a quote from another dealer to bargain, and being obstinate about the price he or she was willing to pay. Often the customer's spouse or friend suggested checking out another manufacturer's offerings, reminded the customer of another vehicle he or she had considered elsewhere, or advised the customer to take more time to make up his or her mind.

In many cases, the interaction between customers and salesmen became strained. It was not uncommon for one party to insult the other. Many negotiations, therefore, never reached an agreed-upon price and, hence, a deal. However, if a deal was struck, the atmosphere became less tense and the customer moved on to completing the paperwork on the purchase.

Internet Sales

During the mid-1990s Internet sites began to emerge (the first was Autobytel) that offered car buyers an alternative to shopping for

---

[27] See Barley (2015) for detailed examples of moves and interactions during negotiations and during earlier phases of the script.

cars on dealers' lots. The first such sites were run by third parties that allowed customers to search for free for cars that they wanted and provided an estimate of a dealer's invoice price as well as MSRP. Soon after, affinity groups, such as Costco, began offering their members similar services and went a step further by negotiating guaranteed prices with local dealers. Manufacturers also began to sell cars through the Internet, although they did not provide information on invoice price or negotiate guaranteed prices. By the time of our study (2007–8), most dealers also had their own website, where they displayed their inventories. In all cases, however, if a customer was interested in buying a car, the website would refer the customer to a particular dealer by emailing the dealer about the customer and his or her interests. In response to these Internet sites, dealers began to set up Internet operations to handle online customers. Typically, the Internet sales office was staffed by different salesmen than those who worked the floor. In both of our sites the Internet offices were located backstage away from the showroom.

As displayed in Figure 2.8, encounters between Internet salesmen and customers followed a radically different script than did floor sales. The encounter consisted of two acts that often occurred on separate days. The first occurred on the phone and the second face to face. An email from a customer or a request for information routed to the dealer by a website initiated the first phase of the encounter. In response a salesman called the customer and/or sent the customer an email, often containing a price quote. Regardless of the medium of initial contact, the salesman's objective was to speak with the customer by phone. If the salesman successfully reached the customer and the customer was still interested, the salesman introduced himself and reiterated the customer's request. Using databases as a guide, the salesman told the customer what vehicles the dealer had in inventory and, if necessary, to what vehicles the dealer had access through other dealers.

If the customer expressed interest in one of the vehicles, the salesman would quote the customer a price. Specifically, the salesman would tell the customer the dealer's invoice price on the car and then state that the dealer wanted an additional sum, representing the dealer's profit on the sale. The invoice price that the

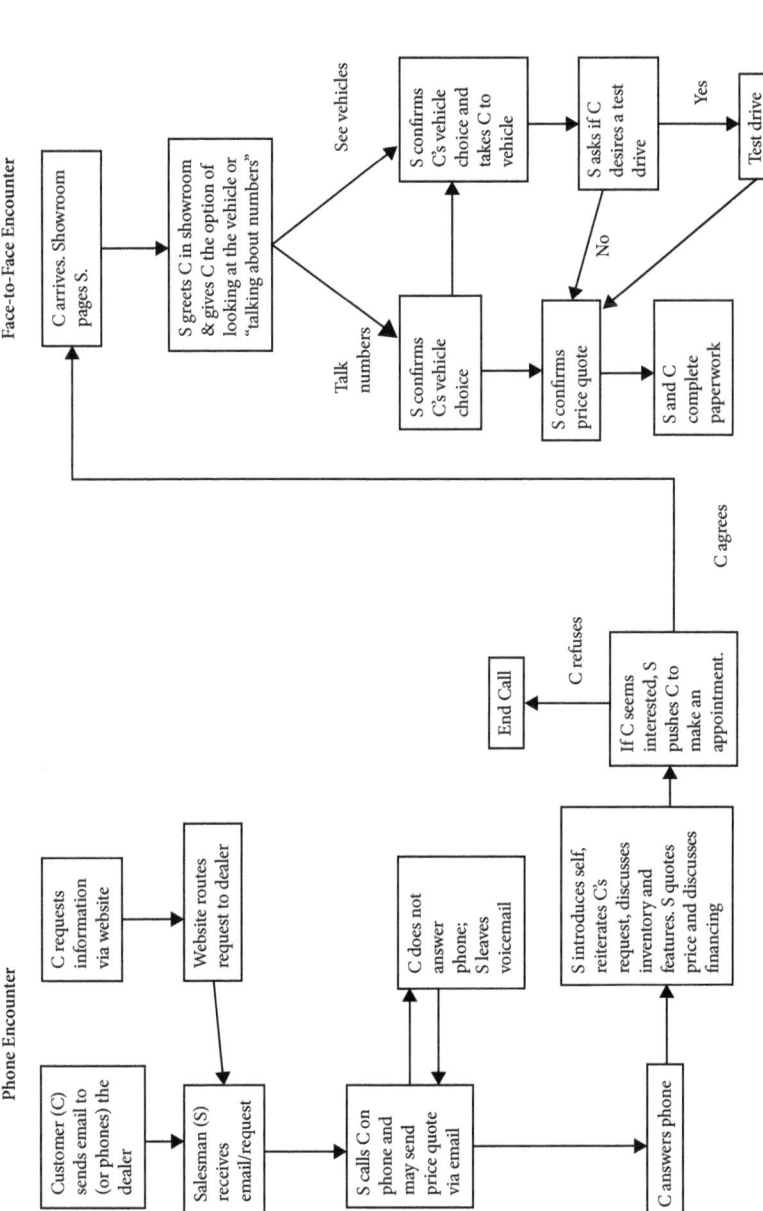

Figure 2.8. The Internet Script

salesman quoted was always accurate, and the profit might be no more than a few hundred dollars, especially if the model was relatively inexpensive. In fact, Internet salesmen sometimes quoted prices under invoice if the dealer wanted to move a vehicle out of inventory. After discussing price, the Internet salesman broached the topic of rebates or financing, giving the customer a choice between the two.

In Internet sales, the quoted price was final. Consequently, haggling did not occur and the Internet salesmen had no need for any of the moves for which floor salesmen were notorious. In fact, the only common move was for Internet salesmen to question prices that customers reported being quoted by other dealers. Typically, the salesman was right and the customer was wrong (usually because the two were talking about different models), and on several occasions when the customer was correct, the salesmen actually suggested that the customer buy the vehicle from the other dealer. If the customer was not interested in the price the salesman offered, the interaction ended. But if the customer continued to be interested, the salesman pushed the customer to make an appointment to come to the dealership, meet the salesman, and test-drive the car.

Agreeing to an appointment was the transition to the second act of the script. Customers who agreed to an appointment arrived at the showroom and told one of the floor salesmen or a clerk that they had an appointment with the Internet salesman they had spoken with on the phone. The floor then paged the Internet salesman, who came to the floor, greeted the customer, and gave the customer a choice of looking at the vehicle or going to the salesman's cubicle to "talk numbers." In the first case, the salesman took the customer to the vehicle, went over its features, and asked the customer if he or she would like to take the vehicle for a test drive. We never observed an Internet customer leave the dealership after a test drive as we did on the floor. If the customer did not want to test-drive the vehicle (or once the test drive was over), the salesman led the customer to his cubicle. Once they were seated, the salesman confirmed the quoted price, often turning the computer screen toward the customer so he or she could see the database from which the salesman was reading. At this point, Internet customers typically agreed to buy the car, and the salesman and the customer proceeded to complete the paperwork. Our data shows that Internet customers were generally more satisfied with their experience than were floor customers. As one put it, the Internet salesmen "weren't pushy."

In short, the Internet script transformed salesmen and their customers from contentious bargainers into price givers and price takers. At both dealerships an Internet salesman was typically the top salesman of the week. As occurred in the hospitals, the Internet bifurcated the two dealerships. Floor salesmen and Internet salesmen were located in physical spaces distant from each other. Floor customers were, therefore, unable to observe interactions between Internet salesmen and their customers and, hence, could not learn how different the two ways of selling were. The bifurcation in the two car dealerships reflected the dealers' decisions. The dealers did not want Internet encounters to contaminate floor encounters. Floor salesmen would have had difficulty successfully playing out a floor script with customers struggling to whittle away at MSRP, if they could observe other buyers being offered a set markup over invoice by salesmen who did not use pressure tactics. Furthermore, it would have been difficult for the dealers to keep their floor salesmen motivated if they could observe Internet salesmen closing more deals than they did. In short, without segregation, dealers could not have maintained, as Goffman (1961a: 26) put it, "a world of meaning exclusive" to the floor.

Why did the Internet alter the role relationship between customer and salesman? The Internet unanchored the sales encounter from its historical mooring in face-to-face interaction, thereby demanding a reconfiguration of interaction during a sales encounter. The reconfiguration changed the definition of the situation in ways that required salesmen and allowed customers to play their roles differently. During the first phase of the encounter customer and salesman shared no common stage. Initial contact occurred entirely over the phone, which enabled the customer to more easily terminate the encounter. If necessary, he or she could simply hang up on the salesman. Unless a customer made an appointment, the salesman could not take advantage of the props, particularly the vehicles that were so central to the floor script. Nor could Internet salesmen avail themselves of supporting actors to create pressure on the customer to buy. Instead, the Internet salesmen had to work entirely with information contained in databases. Under these conditions it would be disadvantageous for a salesman to misrepresent the data, because doing so would eventually undermine the sale if the customer arrived only to find that the deal had changed. Thus, the Internet pushed the salesman to be highly factual and to forgo the stance of a negotiator to sell vehicles successfully.

The stance the salesmen assumed and the script they enacted during the phone encounter carried over into subsequent face-to-face encounters with customers. With the price settled, salesmen had no need to establish rapport. They no longer needed to pressure customers, because customers usually arrived on the lot committed to buying as long as the price remained constant, they qualified for financing, and the vehicle performed as expected. As a result, Internet salesmen had no need for the armory of moves that floor salesmen have devised over the years to persuade customers to buy.

To put it differently, the Internet transformed the selling of cars into a volume business rather than a business based on maximizing margins and commissions. In sharp contrast to the research on distributed teams which indicates that relationships experience more trouble when people must interact with those who are not co-present, in the case of car sales, distance improved the tenor of relationships between customers and salesman.

Discussion

I have found encounters, roles, role relationships, and networks to be valuable conceptual primitives for studying the organizational implications of technological change for several reasons which I have tried to demonstrate with my extended examples from radiology and automobile sales. First, it is worth noting that these concepts have long been the fundamental building blocks of the analysis of social order. They thereby allow us to link the study of technological change to a wealth of previous social research and theory. Second, focusing on what people do when they use a technology, with whom they interact, and how those interactions unfold allows researchers to document relatively rigorously as streams of behavior what others have called practices (Boczkowski and Orlikowski, 2004) or routines (Feldman and Pentland, 2003). In general, if technologies occasion no change in how people do things and how they interact with others, then almost by definition a technology's implications for the social order are minor. Jet engines allow us to fly to places more quickly and require longer runways, but otherwise they have not changed the social organization of air travel in ways that did not already exist with prop planes. (Of course, the airlines' desire to turn a profit has dramatically altered the flying experience, from the lack of free food in economy class to a decrease in leg room and charges for checked baggage.)

Third, a role-based approach enables analysts to document how and why technologies occasion organizational change from the ground up. The goal is to move from actions to interactions to social structure in a more tightly integrated empirical and analytic way. Thus, the approach inverts that of contingency theories of technology and organizing (Woodward, 1958; Hage and Aiken, 1969; Gerwin, 1979; Fry, 1982), which move from the outside in: from the properties of an environment or a production system to a prediction of how the social organization of work will be structured. Contingency theories breed determinism because of their level of analysis and because their dependent and independent variables are often conceptually correlated (see Barley 1986 for a discussion of this problem). In contrast, role-based studies leave many possibilities open, including the possibility that the same technology will affect organizations differently without shutting off the discovery that there may be common outcomes and general patterns of change. With a role-based approach, commonalities across organizations are grounded in human activities and must be empirically demonstrable. Theoretical propositions about outcomes are largely irrelevant. What matters is whether similar patterns occur empirically in multiple settings. At best, emphasizing roles, role relationships, encounters, and networks informs a theory of process rather than a theory of outcomes.

Finally, a role-based approach naturally integrates the social and the material. Studying roles allows us to see clearly that, as Brian Pentland put it, "people use tools to do tasks."[28] Technologies shape what we do and how we do it. Actions, technologies, and even interpretations are tightly entwined not simply in an ontological but a practical sense of the term. Technologies figure prominently in the scripted encounters that lend structure to our goings-on. They are among the material resources that we enroll in the process of doing things. They are no more or less important than the social resources we employ in the same doing. Though potentially simple-minded, the analytic approach I have sketched removes the mystery of the distinction between the social and material. Both represent constraints on and opportunities for our actions, and both contribute to definitions of situations that allow us to get on with our lives and work.

By arguing that the material and the social are fused in ongoing action, I do not mean to preclude the notion that the arrangements I have described were not socially constructed. As I pointed out in my paper on how CT operations became structured at the two hospitals (Barley, 1986), the nature of interactions between radiologists and technologies emerged through ongoing encounters. The radiologists could have chosen to do sonography themselves, as do many obstetricians, but doing so would have required significant changes to the roles of the radiologists. Similarly, Saturn dealers employed what was essentially the Internet script to sell cars long before the Web was widely available. In fact, Saturn was known for its "no haggling" approach to pricing. Clearly, the social orders I have described were socially constructed. The technologies did not make them come into existence. Yet what no choice or action could alter was the cybernetic nature of sonography, the decidedly non-cybernetic nature of X-rays, and the absence of co-presence when shopping on the Internet and talking on the phone. These are material constraints or affordances (depending on your point of view) with which or around which social construction has to deal. Hence, the material implicates the social and the social implicates the material.

---

[28] See Doctor Decade (Pentland's alias) singing "Sociomateriality" at https://www. youtube.com/watch?v=68cEbf5CaBo, accessed December 30, 2019.

Medical imaging examinations and car sales are classic cases of encounters. They have identifiable beginnings and endings, and within that time span they unfold as people go about playing roles according to fairly well-defined scripts on distinctive stages with particular props. Therefore, one might argue, with appropriate comparative or longitudinal data, it is relatively easy in well-defined encounters to observe how technologies affect roles, role relations, scripts, stages, props, and supporting actors to create new definitions of the situation. One might therefore ask, "Does this mean that the utility of the role-based approach is confined to situations in which technologies play a part in encounters that have discrete beginnings and discrete ends? What happens when work flows cannot be so easily chunked?"

There is no doubt that clearly demarked encounters make role analysis easier to do. But the approach's utility is not confined to simple and highly structured repetitive encounters. A role-analytic approach to the study of technology can be also be applied in much more complex and fluid contexts. For example, Diane Bailey and I used a role-based approach when we extracted teaching-learning episodes from the field notes on the complex and less well-scripted work of structural engineers and chip designers (Bailey and Barley, 2011) We were able to show that because of the different types of technology that these engineers employed, the relative rates of change in the two fields, and the different market and institutional pressures under which each labored, the senior, mid-level, and junior engineers in the two occupations played very different roles in the provision of information and help. In structural engineering senior engineers did the bulk of the teaching, whereas junior engineers did the bulk of the learning, and their role relations encoded a network with a strict hierarchy. In contrast, in chip design all engineers taught and learned from each other regardless of tenure. The structures of their role relations encoded a completely connected network in which each role was structurally equivalent.

Paul Leonardi, Diane Bailey, and I also employed a role-based approach to understand how computer simulation technology had altered the work of engineers in automobile design and testing

(Bailey, Leonardi, and Barley, 2012). The study involved not only participant observation, but the use of retrospective interviews and archival documents because we wished to speak to a history of technologies employed in automobile design and testing. Because these technologies spanned roughly four decades of change, we were only able to document the most recent (the use and outsourcing of Finite Element Analysis) using ethnographic methods. Nevertheless, we found that the firm's successive adoption of CAD, computer-aided simulation, and ultimately the use of computer-mediated communication technologies to offshore aspects of design and testing traced a history of changed positions, roles, and interactions. As the automobile firm adopted new technologies, upper-level managers reorganized the work by eliminating some positions and creating others. For example, as CAD became integral to design work, the company no longer felt it need drafters, who had created blueprints for the parts designed by parts designers. When parts designers used CAD systems to design parts, the systems automatically generated the digital equivalent of blueprints. Accordingly, the company fired all drafters and restructured the parts designer's role into what became known as that of a "design engineer." Eventually, managers also created a new role known as "simulation engineer." The changes initially altered the non-relational and relational roles of the parts designers and then the design engineers. Once the company began to outsource the creation of simulation models to India, the company's engineering centers set up different systems for managing the workflow between the centers and India. The American center employed gatekeepers to mediate between local simulation engineers and those in India, whereas the Mexican center let its simulation engineers interact directly with the Indian engineers. As Leonardi and Bailey (2008) showed, role relations and practices at the two centers varied dramatically, and the variation had significant implications for productivity as well as the accuracy of the simulations.

As a final example of the utility of a role-based approach to understanding technological change, consider Schultze and Orlikowski's (2004) study of an online quoting system deployed by WebGA, a firm that assisted independent insurance agents in determining which carriers offered the best policies and prices for the clients the agents served. The online quoting system allowed agents to generate quotes themselves, thereby bypassing WebGA's service representatives, who had historically provided quotes and acted as consultants to agents putting together proposals. By splitting the activities of quoting and consulting and by automating the former, the technology decreased the frequency and altered the tenor of interactions between sales representatives and agents. Drawing on Gould and Fernandez's (1989) typology of broker relationships, Schultze and Orlikowski argued that the online system transformed sales representatives from gatekeepers who aligned themselves with agents into liaisons who were aligned with neither agents nor carriers. Ironically, the structural change undercut the benefit that agents once gained by working with WebGA.

Aside from noting that the radiologists and administrators who ran the two radiology departments created new positions for technologists who operated the computerized modalities and automobile dealers created distinct positions for Internet sales personnel, I have not directly dealt with power dynamics during instances of technological change. The roles and role relations I have discussed largely emerged in the course of the technologies' use. Nevertheless, it is important to recognize that roles and role relations can be imposed by more powerful actors, as occurred in the case of the automobile company.

It is also worth remembering that technologies do not necessarily occasion changes in role relations. But even when role relations remain intact, technologies can have a significant effect on how work is done and what one needs to know in order to do a particular line of work. It is highly unlikely that universities would hire a person for the position of an administrative assistant if the individual possessed only the skills that administrative assistants possessed forty years ago. Similarly, the work of orthopedic surgeons has changed dramatically as surgical technologies and medical devices like artificial knees have evolved. Yet the role of the orthopedic surgeon remains largely what it was thirty years ago. Changes such as these are important. The skills now required of administrative assistants allow universities to operate in a digital world. Changes in how orthopedic surgeons do surgery have allowed patients to recover more quickly and experience better results. My point has simply been that technologies will not occasion change in organizations unless they alter role relations and by extension the networks that define an organization's structure. It is with organizational change and changes in the division of labor, and not simply changes in technology, that my work has largely been concerned.

3

How Should We Study Intelligent Technologies' Implications for Work and Employment?

Stephen R. Barley and Matthew I. Beane

Having laid out in Chapter 2 how technologies change work by changing roles, we return to the issues raised at the beginning and end of Chapter 1. What are we to make of the growing buzz and mounting fears surrounding the anticipated proliferation of intelligent technologies such as machine learning and robotics? Do we really stand on the precipice of a brave new world that will usher in the twenty-first century's equivalent of another industrial revolution replete with hardships and perhaps new opportunities? Will jobs currently performed by humans, especially jobs requiring complex knowledge and the control of production, yield to the inanimate and go the way of blacksmiths, wheelwrights, and other jobs prominent before the Second Industrial Revolution? Is our future one of even greater underemployment and unemployment than we face today? Or, as in the past, will new forms of work arise from left field to substitute for work that intelligent technologies will make unnecessary or redundant? Should we fear intelligent technologies or embrace them?

Ultimately, the answers to these questions rest on the agendas held and actions taken by firms, technologists, users, policymakers, and other stakeholders (Bailey and Barley, forthcoming). For this reason, it is difficult to answer such questions with certainty. Variation in agendas, actions, and technical designs as well as

Work and Technological Change. Stephen R. Barley, Oxford University Press (2020). © Stephen R. Barley and Matthew I. Beane. DOI: 10.1093/oso/9780198795209.003.0003 differences in research methods make it impossible to find a single, solid answer to any of these questions. Nevertheless, decision makers must envision some scenario to develop plans that may affect millions of people on critical matters such as employment and immigration policy, tax law, education, and organizational design. Many decades of careful research overwhelmingly suggest that any current certainty will be seriously misplaced. For example, many predicted that making music digital and shareable via the Internet would greatly decrease demand and destroy many jobs in the music industry. Yet studies indicate that more frictionless, subscription-based access to music has increased music consumption and led to a boon: steadily increasing paid attendance at live music events with concomitant gains in employment for musicians and event production staff (Nielsen Company, LLC, 2018; Wolfson, 2018). As noted in Chapter 1, few of us are prescient enough to anticipate the second-order effects of new technologies, especially infrastructural ones. Even the first-order or intended consequences of a new technology are more elusive than we think. The final word on intelligent technologies can only be written by our descendants, who will have the luxury of looking back at history, and, even then, there will be debate. Yet we believe that scholars can do a better job of systematically collecting and analyzing data that will assist us in more clearly and carefully addressing the questions raised in the previous paragraph.

Good empiricism is particularly important, because the more uncertain the times, the more the world turns to scholars, researchers, and public analysts for guidance. In this chapter we see our primary audience as scholars, researchers, and public intellectuals who study the social and economic implications of intelligent technologies. Accordingly, we first review what thinkers have had to say to date. We then raise two conceptual and empirical problems that hamper current approaches to divining the possible implications of intelligent technologies: an isolationist view of technology and a reductionist view of work. Afterward, we illustrate what addressing these problems would involve, first in the historical case of the railroads as an initial illustration and then in the case of an intelligent technology. In each case, we offer two new conceptual frameworks to guide analysis, the first of which, to our knowledge, has never been used to study technology, work, and employment. The first relies on the metaphor of a "stack" to explicate the hierarchically organized technical substrates that make all focal technologies possible. The second is a model for role systems-driven analysis of how technologies are integrated into workplaces. Finally, we speculate on approaches to research that make use of these frameworks.

Current Thinking on Intelligent Technology, Work, and Employment

The current round of discourse on how intelligent technologies will influence jobs and employment began around the mid- to late 2000s as economists sought reasons for wage polarization and the decline of well-paid jobs in manufacturing and clerical work (Autor, 2007; Autor, Katz, and Kearney, 2008; Acemoglu and Autor, 2011). In 2011 Brynjolfsson and McAfee published the widely read book, *Race against the Machine: How the Digital Revolution Is Accelerating Innovation, Driving Productivity, and Irreversibly Transforming Employment and the Economy*. The authors cogently summarized the largely macroeconomic research on the implications of intelligent machines for work and the economy. They then extrapolated underlying technical trends and explored the implications of the world they envisioned. In particular, Brynjolfsson and McAfee suggested that the jobless growth which seemed to have occurred after the two most recent recessions might be attributable to the spread of computer control and intelligent technologies.

Within five years a spate of books appeared on how artificial intelligence (AI), robotics, and other intelligent technologies could alter the structure of work and employment (e.g., Brynjolfsson and McAfee, 2014; Ford, 2015; Kaplan, 2015; Markoff, 2015; Mindell, 2015; Domingos, 2015; Susskind and Susskind, 2015; McChesney and Nichols, 2016; O'Neil, 2016; Schwab, 2016). During the same period, national and international bodies took up the topic from a policy perspective (World Economic Forum, 2016; Furman et al., 2016; National Science and Technology Council, 2016a, 2016b; National Academies of Science, Engineering and Medicine, 2017). Consulting companies began issuing related reports and forecasts (e.g., Chui, Manyika, and Miremadi, 2015; De Smet, Lund, and Schaninger, 2016; Bughin, Lund, and Remes, 2016; Muro, Maxim, and Whiton, 2019). As the volume of analysis increased, the debate quickly filtered into the popular media (newspapers, magazines, blogs, podcasts, television, and the like) to inform lay discussions of AI, robots, and employment.

Many of the books and government reports captured attention and spurred debate by building their arguments around tales of the wondrous capabilities and achievements of specific robots, specific implementations of machine learning, the spread of sensors, or all three in some combination. Common examples included the defeat of chess champion Gary Kasparov by IBM's Deep Blue (Wikipedia, 2019), IBM's Watson's triumph over Jeopardy champions Ken Jennings and Brad Rutter (Wikipedia, 2020), Watson's subsequent deployment to Memorial Sloan Kettering Cancer Center to help diagnose lung cancer (Memorial Sloan Kettering, 2012), driverless cars developed by Waymo, Amazon's deployment of Kiva's (now Amazon Robotics') robots, which transport items from warehouse shelves to the humans who pack shipments (Healer, 2019), and the intelligent platforms that deploy and direct Uber or Lyft drivers (Calo and Rosenblat, 2017). Authors then extrapolated from such examples to the dilemmas that intelligent technologies supposedly pose for the economy, work, and employment, even though some of the examples (e.g., chess and Jeopardy) had little to do with workplaces or employment.

Different authors made different assumptions about the workings of capitalist economies, about the technologies, about the technologies' paths of development, and about the visions and motives of those who design and adopt intelligent technologies. As a result, the literature clusters around three general perspectives on the future of work and employment: optimistic, pessimistic, and more nuanced views. For example, by beginning their argument with the historical fact that previous industrial revolutions eliminated entire occupations while creating new lines of work, Brynjolfsson and McAfee (2014) ultimately arrived at an optimistic view of the future, while readily admitting that during the transition many people would experience hard times. Ford (2015), on the other hand, emphasized the concept of technological unemployment to arrive at a more pessimistic outlook on the future of work and social life. In contrast, taking a more nuanced view, Markoff (2015) distinguished between two long standing perspectives among computer scientists, characterized by the contrast between John McCarthy's and Douglas Engelbart's distinct philosophies of computer design. Markoff argued that intelligent technologies can either be designed to replace human workers or to augment human abilities and the work people do. Mindell (2015) also offered a more nuanced view. He drew on examples of robots and intelligent technologies in oceanography, aerospace, and the military to point out that such technologies require support systems that employ many humans who enable the technologies to operate in what mistakenly appears to be an autonomous fashion. It is difficult to determine whose visions are more persuasive because most authors are forced to spin their tales in the absence of a sufficient body of empirical data that would allow us to make crucial distinctions and identify contingencies that might lead to one outcome or another.

Frey and Osborne's (2013) working paper entitled "The Future of Employment: How Susceptible Are Jobs to Computerization" was an early and much-cited milestone in the quest to estimate empirically the impact of advanced automation on employment. Frey and Osbourne used the task descriptions of 702 jobs listed in O*Net, the US Department of Labor's replacement for the *Dictionary of Occupational Titles*. Operationally, they had experts in machine learning estimate whether seventy occupations were susceptible to automation, assuming the current state of intelligent technologies. They used these experts' assessments as training data to build a machine-learning algorithm that assessed the susceptibility of the remaining 632 occupations to automation. The algorithm's results suggested that 47 percent of the occupations in the United States were *susceptible*.

Although Frey and Osborne were careful to point out that they were not saying that 47 percent of occupations *would be* replaced by machines, their figures were frequently interpreted as a strong prediction, especially in the popular media. Frey and Osborne's work triggered additional studies that also attempted to arrive at estimates (McKinsey Global Institute, 2017a, 2017b, 2018; Organisation for Economic Co-operation and Development, 2016). Using three different scenarios that varied by speed of adoption, McKinsey (2017a) estimated that up to 30 percent of work activities worldwide could be affected by robots and automation by 2030. However, McKinsey was not transparent about how they constructed their data or models. The OECD's research was more methodologically transparent, but in contrast to Frey and Osborne, they estimated that only 9 percent of jobs in the US were subject to elimination by automation.

How can researchers arrive at such vastly different estimates? The answer depends, in part, on the methods they use and, in part, on the unit of observation they adopt. Frey and Osborne used occupations as their unit observation, McKinsey used work activities and skills, and the OECD focused on tasks. Our hunch is that tasks or work activities are the most reasonable of the three, because jobs often entail multiple tasks and because deploying technologies can alter the nature of the tasks people perform without eliminating their jobs or their occupation. Of course, if a technology can automate or substitute for the majority of the tasks performed by members of an occupation, then the occupation is likely to disappear, as occurred when computer-aided design (CAD) systems gradually eliminated the need for drafters (Bailey, Leonardi, and Barley, 2012).

Our approach to conceptualizing how intelligent technologies may alter work and employment differs from most previous writing in two ways. First, we do not subscribe to the view that intelligent technologies merit being treated as a trigger for an industrial or technological revolution. As discussed in Chapter 1, technological revolutions rest on a small set of infrastructural technologies that trigger the swarming of new technologies which gradually come to alter society's system of production. In our view, computerization, or perhaps more accurately digitization, has been shifting the technical infrastructure of at least Western societies for over half a century. Intelligent technologies certainly extend the process of digitization because they rest and build on a digital base that is already in place. In other words, talk of a Fourth Industrial Revolution (Schwab, 2016) is a misconception, although it may serve certain actors (both individual and organizational) as a useful rhetorical ploy. If we are experiencing a new industrial revolution, it would be better to speak of a digital revolution and treat intelligent technologies as a key development in the swarming of technologies that has accompanied all past technological revolutions.

Second, as articulated in Chapter 2, we assume that no technology can be said to change jobs, work, or employment unless it reorganizes work practices, roles, role relationships, and, ultimately, the networks that constitute local structures of production. Although some students of intelligent technology occasionally mention infrastructure (Autor, 2019) or role relations (Faraj, Pachidi, and Sayegh, 2018), few have used either notion to discipline or organize their empirical or theoretical approach to assessing implications for work and employment.

Instead, most of the literature on intelligent technologies makes the common, although often implicit assumption that intelligent technologies are somehow independent of a larger technological system. For lack of a better term, we will call this an *isolationist view of technology*, one that divorces a technology from the technical substrate on which it rests. By a technical substrate we mean the array of technologies and work systems that enable a specific technology to exist and operate.

The tendency to treat technologies as isolated artifacts is not confined to researchers and commentators on technological change. Most of us treat technologies as if they were independent of the technological systems that allow them to function.[1] Hence, we talk

---

[1] With the exception of scholars associated with the social construction of technology (Bijker and Pinch , 1987), historians of technology are far more likely than most social scientists to attend to the material and social systems that produce technologies and to how technical systems change. Most historians describe rather than explain or predict, however. Thus, the perspective of the historian is informative for the kind of analysis that we propose, especially for the analysis of technical stacks.

about toasters as if they were not reliant on the electrical grid; about knives and forks as if they were independent of metallurgy, mining, and refining; and about cars as if they were independent of the hydrocarbon infrastructure or the rules, roads, signs, and signals that allow people to operate cars in an orderly and reasonably predicable fashion. Note that in each case, the work created or changed by adopting a toaster, a fork, or a car occurs not simply among those who use the toasters, forks, or cars, but also among those who create and maintain the components of the technical substrate on which each rests. Analysts typically either miss or fail to address this point when pondering the implications of intelligent technologies for work and employment.

Furthermore, most current commentary on intelligent technologies unfolds as if there was a direct link between employment and the automation of a skill, task, or job. In other words, analysts treat jobs as if they were unconnected, even though work typically unfolds through an interwoven set of social relations. For lack of a better term, we call viewing jobs as isolated entities a *reductionist view of work*. We use the term to imply that while tasks and skills are important to understanding the implications of intelligent systems, they represent only one of several interdependent units of analysis. Overlooking technological interdependence, analysts see that a robot can weld two parts together and then draw conclusions about the automation of welders' work. Others see inexpensive tax preparation software hit the mass market and presume that tax preparers will disappear (for why they have not, see Galperin, 2017). Conversely, some commentators look at intelligent technologies and predict new jobs required to manage them (National Academies of Science, Engineering and Medicine, 2017).

The important point is that changes in employment *only potentially begin* with changes at the level of a task. Changing a task may or may not eliminate a job, may or may not require substantially new ways of working, may or may not create the foundation for a new occupation, and may or may not occasion changes in other jobs and roles. In short, reductionists treat tasks and jobs as if they were neatly bounded activities that can be replaced or complemented by technology. But, as noted in Chapter 2, tasks are embedded in and constituent of work practices—there are few tasks performed without tools and none that is devoid of context. Work practices are, in turn, attached to roles. Roles are embedded in and constitutive of role relations that define networks of roles. Change at each of these levels of analysis is negotiated across jurisdictional boundaries, involves practical difficulties and opportunities, and exhibits the dynamics of a complex social system. A reductionist view of work blocks inquiry into these interdependent domains, making it difficult to build a more encompassing account of how a technology will affect work and employment.

Framed positively, our claim is that changes in work and employment associated with an intelligent technology are likely to propagate "vertically" throughout its technical substrate and "horizontally" across role systems, that is, technologies are technologically embedded and their use is embedded in a role system. How propagation happens will vary by the specific technology and by the settings in which it is used. We, therefore, argue that assessments of an intelligent technology's social impact should be traced both vertically (up and down technical substrates) and horizontally (across role systems).[2]

Why are isolationist and reductionist views so common? On the one hand, the tendency toward rationalization of work-related phenomena has long tended to favor simple models of cause and effect. For example, Taylor (1911) insisted that productivity was primarily a function of the tools and methods that workers used and that both could be optimized by studying a limited number of factors, such as tool size, fatigue, optimal sequences of motion, and so on. This impulse to rationalize has produced theories and models that guide the development of science and technology. We shall focus here on two additional issues: isolationist and reductionist views of technology and technological change. These views often flow from a lack of familiarity with the technologies themselves as well as a

---

[2] To write this paper, we read seventy-six academic articles written by economists, historians, sociologists, and students of science, technology, and society. We read an additional 148 articles, books and reports in the popular, policy and consulting literatures. We found no evidence that any author adopted a technologically embedded view of intelligent technologies, and very few who have adopted a role systems view.

lack of direct longitudinal and contextual data on how technologies are deployed and used in practice. Analysts' conclusions are bound to be incomplete, if not inaccurate, without both kinds of knowledge.

Not understanding a technology can lead analysts to make three unwarranted leaps. The first involves mistaking a toy for a tool. Many analysts move from the existence of a prototype to arguing that it is now possible to automate a line of work. The prototype serves as an icon or a symbol for an imagined future that does not yet exist. Demonstration videos, such as Microsoft's personal assistant mentioned in Chapter 1, make it look as if the system can automate work that was previously within the "bright red line" of what only humans can do.[3] But when examined closely, the system's functionality is quite narrow and limited to a highly structured environment or situation. In the case of the personal assistant, the fiction is that a personal assistant's only tasks are to manage someone's calendar and take messages.

Another problem with not understanding a technology is that it becomes easier to assume that the technology's influence will be the same across contexts. In this sense the popular and academic literatures are similar, with the popular literature going to greater extremes. Lacking sufficient technical knowledge, commentators make incorrect assumptions about how technologies such as machine learning or bin-picking robot arms function. Susskind and Susskind's (2015) work is an exemplar. Their thesis is that AI combined with the Internet and mobile computing will make the expertise of most—if not all—professionals widely available at low cost. On these grounds, Susskind and Susskind predict the demise of the professions. Making such a leap requires overlooking the fact that most intelligent systems are built to do one thing or make a narrow range of predictions. In contrast, professionals typically handle a variety of problems often marked by relevant and sometimes subtle nuances. It is extremely difficult to build machine-learning programs that can handle a variety of situations and

---

[3] Eric Horvitz's TED talk shows the personal assistant in action, https://www.youtube.com/watch?v=dpoVh9xwdD4, accessed December 29, 2019.

recognize uncommon but crucial nuances. Moreover, Susskind and Susskind overlook the inconvenient social fact that professional dominance rests on laws that dictate who can give professional advice and who is liable when the advice is wrong. Last, but not least, Susskind and Susskind downplay the power of professions to resist people and machines that challenge their occupational jurisdictions. For both technical and social reasons, AI will not cause professions to go as quietly and as inevitably into the night as Susskind and Susskind seem to believe (see Abbott, 1988, Zetka, 2003).

Failing to collect direct contextual and longitudinal data on how a technology is deployed and used can also foster empirically mistaken claims. All too often commentators base their arguments on a single site visit, a handful of interviews, or the story of an individual's experience with a new technology. Although this problem is more common in popular and journalistic accounts of intelligent technologies, it can also be found in scholarly writings. Moreover, analysts sometimes uncritically accept the predictions of those who design or sell the technologies. With such thin data we do not know whether conclusions drawn based on such observations, experiences, or claims are representative. Worse yet, we often do not have enough information to determine whether the analysts' interpretations are warranted. The result is often a well-written but poorly substantiated storyline that fosters either fear or hope. Empirical elision is self-sealing: It deprives analysts of the opportunity to build a vivid, contextualized, and contingent understanding of the ways in which change actually propagates through work systems such as organizations.

To illustrate how analysts might begin to overcome isolationist views of technology and reductionist views of work systems, we turn first to the coming of the railroads and then to AI. We begin with railroads for two reasons. First, railroads are familiar to most readers. Second, our stance is that all technologies are embedded in a technical stack as well as a role system. Thus, we should be able to approach an analysis of any technology's implications for work and employment similarly.

A Technologically Embedded and Role Systems View of Railroads

Technological Embeddedness

As defined in Chapter 1, rail systems count as an infrastructural technology: They altered systems of production in the US, Europe, and elsewhere. Numerous scholars have written exacting and insightful analyses of railways and the profound impact they had on social, economic, political, legal, and technological institutions (Hadley, 1892; Stover, 1961; Chandler, 1977; Deverell, 1994; Daniels, 2000). Railroads represented a new form of transportation and shipping—a transfer system, writ large, in Faunce's terms. They not only made land travel faster than wagons and canals, but they made possible trade between cities and other locations that were previously isolated. With the coming of the railroads, cities in the hinterlands of the US grew, and producers were able to sell their goods in more distant markets, which enabled the expansion of production, the advent of the first large corporations that required coordination, and the rise of management as an occupation (Chandler, 1977). The Baltimore and Ohio (B&O) Railroad began constructing its first track on July 4, 1828, which connected Baltimore to Sandy Hook, Maryland. By 1883 four transcontinental railroads were completed. By the end of the nineteenth century, railroads had become essential to commerce in the US and elsewhere: Railroad-related employment in the US hit 749,000 in 1890 and peaked at 2,076,000 in 1920 (U.S. Bureau of the Census, 1975).

A technical isolationist interested in how the railroads occasioned changes in work and employment might approach the topic from two directions. First, they would enumerate the jobs involved in running a railroad and ask how many people held them. For example, they might ask, "How many signaling jobs were created as railways expanded and over what timeframe?" or "What did these jobs pay?" Or, they might consider how railroads were built and then ask, "How many jobs did the building of each transcontinental railroad create and how many of these jobs required unskilled labor, craft labor, and knowledge work?" In search of answers, isolationist researchers would scour historical records for employment data and qualitative descriptions of the work involved in constructing and running early railroads.

Second, technological isolationists might explore employment in those modes of transportation for which the railroads substituted. Prior to the advent of the railroads, the primary means for transporting goods over land was wagons and canals. In fact, the owners of early railroads such as the Baltimore and Ohio intentionally sought to compete directly with canals. As railroads spread, the use of canals gradually declined, and canal workers lost their jobs as a result. Canals involved numerous jobs, including hoggees (boys who guided horses and mules along towpaths to pull barges), barge captains, cooks, and operators of drawbridges and locks. Technological isolationists would ask how rapidly did such jobs decline, did their incumbents find new jobs and, if so, in what industries.

Make no mistake, the questions asked by an isolationist are important, but each takes a narrow view of the technologies that were involved as railways became the dominant mode of transportation and shipping on land. Locomotives, cars, tracks, bridges, and switches were just the tip of the technical iceberg that enabled railways to function. The expansion and increasing complexity of the railroads depended on the rise of other technologies that constituted the technical substrate that made rail travel possible. As early as the middle of the eighteenth century two technologies emerged that greatly facilitated the production of the amount of iron that would be required for locomotives, rails, bridges, railroad spikes, and other components of the rail system. The first was the mass-scale production of coke, a very hot-burning fuel produced by heating coal or oil in the absence of air. The second was the blast furnace, a device that forced superheated air into a smelting chamber at far greater than normal atmospheric pressure.

To analyze a technical substrate, we imagine the substrate as a hierarchy or stack of technologies as well as the techniques, organizations, and labor that produce each technology. The technologies in each layer of the stack draw on or are composed of technologies in the layer below. Thus, we can depict a technical stack as a tree. At the top (or the apex of the tree) is the *focal application*: a technology designed and built to fulfill particular purposes. The second layer consists of what we shall call *platforms*: the interconnected subsystems that make the focal application usable or valuable in practice. Platforms are themselves constructed of *subsystems* (the third layer): a suite of components that execute a relatively focused function. Subsystems are assembled from *components* (the fourth layer): a collection of nonvolatile resources that allow a very wide range of applications to function. Components depend on *refinement* (the fifth layer): making raw materials suitable for the production of components. Finally, refinement processes hinge on *extraction* (the bottom layer): converting natural phenomena into usable raw materials. Figure 3.1 lays out the logic of the various layers of a stack. This framework can be used to chart the technical embeddedness of a technology moving from a focal application backward in the supply chain to the conversion of natural resources into raw materials. The italics in Figure 3.1 define the levels of embeddedness for any technology, including intelligent technologies.

Figure 3.2 illustrates the hierarchically ordered technical stack in which a rail service was embedded. For example, railroads required

Focal Application	A technology designed and built to fulfill a particular purpose for users
	(Any of a variety of technologies)
Platforms	Interconnected subsystems that make an application usable or valuable in practice
	Connected material and/or digital environments
Subsystems	A suite of components that execute a relatively focused function
	Application specific devices and/or software
Components	A collection of nonvolatile resources that allow a wide range of applications to function
	Generic parts and/or system libraries
Refinement	Making discrete phenomena suitable for component construction
	Processing facilities and/or operating system
Extraction	Converting natural phenomena into raw materials
	Mining and harvesting equipment and/or drivers and sensors

Figure 3.1. Framework for Embedded Analysis of Technology

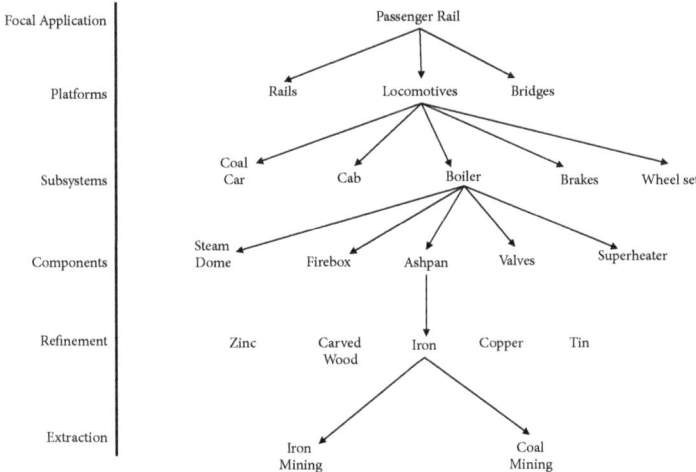

Figure 3.2. Technical Base for Passenger Rail

a number of platforms such as locomotives, cars, tracks, bridges, switches, and stations. Each of these platforms was constructed of numerous subsystems; in the case of a steam locomotive, the subsystems included a coal car, a cab, a boiler, brakes, and wheel sets.[4] All of these were, in turn, comprised of components. The boiler, for example, required a steam dome, a firebox, an ash pan, valves, and a superheater. Each of these components was made of refined raw materials such as zinc, carved wood, iron, copper, and tin (the last two were required to make brass). The boiler was made mostly of iron. Finally, each refined material was extracted from the natural environment by either mining or forestry. Producing iron required both iron ore and coal. Making elements that composed each layer of the technical stack required labor and, hence, was a source of both work and employment.

To develop a full account of how railroads influenced employment, jobs, and work would require analysts to expand their inquiry and data collection beyond direct employment by the railroads and the loss of jobs through substitution to include work and

---

[4] To simplify Figure 3.2 we expand on the components of only one technology in each layer. Furthermore, some layers involve more technologies than we enumerate. Our agenda with this figure and with those associated with our discussion of intelligent technologies in the next section is meant to be illustrative and suggestive rather than exhaustive.

employment throughout the entire technical stack required to make a functioning railroad. Yet, when assessing the railroad's implications for work and employment, economists, historians, sociologists, and other scholars rarely account for developments in the technical stack. Without considering the stack, estimates of any focal application's total influence on work and employment may underestimate the technology's ramifications.

Role Systems

A reductionist view of work is similarly problematic because thinking in terms of tasks and jobs rarely progresses toward an understanding of how a technology triggers changes in work practices and role systems. For example, as American railroads expanded both locally and transcontinentally, rail companies were plagued by collisions and derailments, especially when trains ran in both directions on a single track. Derailments and collusions continued into the twentieth century, when they reached their peak at 36,313 in 1920 (Williams, 2008: 218). To reduce the odds of such disasters, the railways turned early on to train schedules, signaling systems (Nelson, 2019), the standardization of time zones, and, most importantly for our purposes, the creation of two occupations charged with monitoring the position of trains, operating signaling systems, warning engineers of the position of other trains, and performing maintenance on the signaling systems. "Signalmen" were responsible for operating the signaling systems, while "signal maintainers" built and repaired the signaling systems. In practice, many of the same employees could perform both roles, albeit at different times.

Train passage was initially regulated by hand, with signalers stationed at (and between) stations to signal engineers according to preset rules and locally maintained timekeeping devices. Often signalers used lanterns to alert oncoming trains of danger or to reduce speed. Railroads began automating and standardizing signaling systems in the 1870s and 1880s by deploying automatic block signals (mechanical, and then electrical devices) that alerted a railroad engineer to the presence of other trains ahead.[5] Rail companies hoped automated signaling systems would reduce the number of signalmen and maintainers. As Williams (2008) goes to great lengths to show, neither occurred. Instead, with each change in technology signalmen and maintainers learned new skills, ranging from electrical engineering, pneumatics, carpentry, and machining to blacksmithing, welding, sheet metal working, and pipe fitting. They did so by studying on their own, by teaching each other, and by using documents produced by the Brotherhood of Signalmen, a union formed in 1901.

The case of signalmen and maintainers is important for several reasons. First, these occupations demonstrate that automation can change tasks and skills without weakening or eliminating an occupation. In fact, automated signaling apparently broadened and upskilled the signalmen's and maintainers' tasks and skills. Second, the role set of signalmen and maintainers included engineers, conductors, tower operators, dispatchers, and managers. Although historical documents are not detailed enough for us to describe if and how changes in the signalmen's and maintainers' work led to changes in the work of most of these occupations, we do know that how signalmen interacted with locomotive engineers changed over the years. Early signalmen not only attempted to alert engineers to traffic using lanterns, but they also used brass rings to hand engineers papers that documented what lay ahead on the tracks. Signalmen extended the ring on a pole toward engineers who snagged the ring on their arms as the engine passed by, not unlike how people capture brass rings on merry-go-rounds.

Role relations between signalmen and engineers began to shift with the advent of mechanical and then electrical signaling systems. For example, signalmen began communicating with engineers by controlling signals at the beginning and ends of a block using a series of levers situated in a signal house. Signalmen also learned to

---

[5] A block was a specified stretch of track for which a signalman was responsible. Ideally, only one train could travel within a block at a time. Signals were designed to increase the probability that that there would be only one train in a block regardless of its direction of travel.

use the telegraph to communicate with other signalmen working in blocks adjacent to their own. Finally, the fact that the signalman's jurisdiction was challenged by several craft unions, including the International Brotherhood of Electric Workers, suggests that the changing nature of the signalman's work began to encroach on the work of electricians and other craftsmen, hence, spurring jurisdictional disputes between unions associated with the American Federation of Labor (Williams, 2008). Significantly, signalmen still exist on contemporary railroads, although their work and their role relations have changed tremendously as the occupation appropriated and mastered new and more automated technologies, including computerized signaling technologies.[6]

A Technologically Embedded and Role Systems View of Intelligent Technologies

Like railroads or any other technology, intelligent machines not only rest upon a technical stack, but they are also deployed within social systems composed of work practices and roles. Hence, an intelligent technology's implications for employment, work, skill development, and occupational jurisdictions cannot be fully assessed without considering both the technical stack and the role system in which it is embedded. On the technical side, a functioning system of this kind requires a dizzying array of technologies, which are hierarchically ordered and interdependent in practice. For example, OpenAI's DotA AI system (a system based on reinforcement-learning that can outperform the best human experts at *Defense of the Ancients*, a game with optimization and planning dynamics similar to real-world problems such as supply chain management) relies on a distributed software development toolkit, various software libraries, the Linux operating system and hardware drivers, the combination of which is grounded in hardware such as servers, graphics processing units (GPUs), network switches, fiber-optic cables, and input sensors. As with railroads,

---

[6] Note that signaling systems are what Faunce would call a control technology.

the bulk of the impact of deploying these systems at scale may well lie beyond the focal technology and its immediate context. This is, of course, an empirical question, but those who study the implications of intelligent systems for work and employment need to consider a technology's implications for its technical subsystem.

On the role systems side, a tangle of enacted, interdependent work must be done to field an intelligent system, and people occupying differing roles perform this work. How these roles are related is the result of a negotiated order (Strauss, 1978). A change to any role may challenge the status quo within a network of other roles. Thus, to track the implications of an intelligent technology on work, one must engage analytically with local enacted role systems to make defensible claims about the effects of the technology on a network of interconnected tasks, skills, relationships, and so on.

To illustrate the technical stack of an intelligent technology, we will unpack Affdex, an AI algorithm developed and sold by Affectiva.[7] Affdex classifies emotional responses in real time using facial cues and gestures as input. Affectiva promotes Affdex as a tool for applications such as market research and automotive safety. Using Affdex, a firm can analyze video images of people's responses to a product or object to determine how they react emotionally to the stimulus. Stimuli might be as diverse as a device, a software application, a movie, or another person. Thus, systems like Affdex can also be used not only for marketing but in other contexts such as crowd control and border security.[8]

Affdex is classified as AI because it relies on supervised machine learning to augment and improve its ability to classify affective states by extrapolating from a training data set. Programmers create the training data set by having humans label the emotions that people display in a corpus of video recordings. The data set is then divided into two subsets, the first is used to train the algorithm and the second to test how well the algorithm performs. Affdex's learning algorithms analyze the training data, then attempt to classify

---

[7] For more on Affdex, see https://www.affectiva.com/product/affdex-for-market-research/, accessed January 3, 2020.

[8] See https://www.theguardian.com/technology/2019/mar/06/facial-recognition-software-emotional-science, accessed January 3, 2020.

the remaining images correctly. Given that humans have classified images in both data sets, the system can check itself. Programmers deem the system to have learned when it correctly classifies new faces in the second data set using the portion of the data on which the machine has trained. This learning process has no end, because new populations (e.g., ethnicities, ages) and observational conditions (e.g., bad weather, partial faces) are practically inexhaustible.

Suppose a technological isolationist were to focus on the implications of Affdex for employment in border security. Before Affdex (and similar technologies) governments relied on immigration officers to surveil inbound travelers, assess their emotional state, and make immigration and customs decisions based on their assessment. Should this person be searched and detained for questioning or is the person behaving oddly for some other reason? Traditional analysis might ask: How much of border security work involves the assessment of affect and how many border guards could be replaced or complemented by a system that performs the task roughly as well or better than a trained human?

Such questions focus on border work but ignore the work required to field Affdex in the first place. To properly study Affdex's full implications for work and employment, we need to repeat our railroad exercise: enumerate and interrogate the hierarchically ordered set of platforms, subsystems, components, refinement processes, and extraction processes that allow Affdex to function. However, intelligent technologies, like Affdex, differ significantly from traditional mechanical or electrical technologies. The technical stack of a mechanical or electrical technology is primarily rooted in physical or material objects and the techniques that produce them. Affdex and other intelligent technologies are more complicated. They are at once a complex set of material artifacts that make the collection, processing, and output of data possible and set of digital applications that process data. In other words, Affdex (like all information technologies) is composed of two technical stacks: a material stack and a digital stack that work in concert. We treat the two stacks separately only analytically because the material and digital stack are both necessary for an intelligent technology (or for that matter any computational technology) to

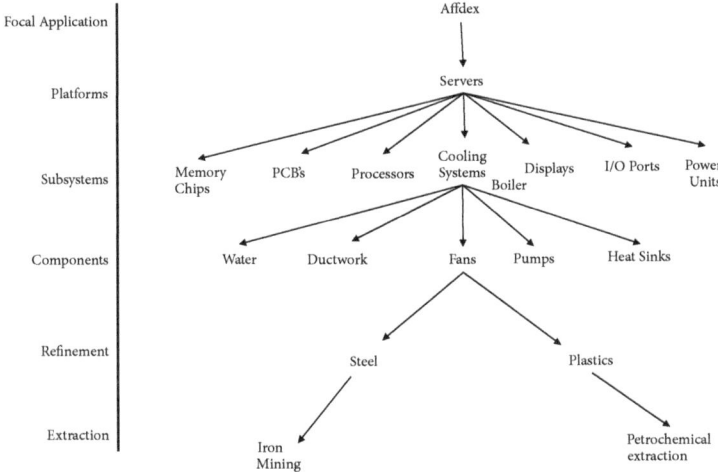

Figure 3.3. Material Technical Base for Affdex

operate. The presence of the digital stack is why commentators view intelligent technologies as pathbreaking developments in the evolving digital revolution.[9] Those interested in how intelligent technologies affect work and employment should explore both stacks. We begin with Affdex's material stack, as depicted in Figure 3.3.

The Material Stack

At the top of the material stack sits Affdex, the focal technology. Computer scientists would call Affdex an *application*, which, like every application, provides a specific and distinct function for those who use it. Applications ingest, process, and analyze often massive data sets to provide insight that, in principle, enables one to better understand and act more effectively in some complex, dynamic domain of work or life.[10] Although applications are typically viewed

---

[9] Of course, one might argue that at the level of electrons the digital is also material. While this is a valid ontological point, it obviates the practical and important distinctions between digital and mechanical technologies.

[10] In this sense, Affdex is no different than Apple Maps, Google Maps, or Waze, which allow drivers to navigate from one destination to another while anticipating traffic slow-downs or perhaps finding alternative routes to avoid them.

as programs or algorithms, it would be impossible for applications to function if they did not "run" on a material stack.

Platforms constitute the highest level of an intelligent technology's material stack. Physically, platforms include the interconnected hardware that undergirds capabilities related to data storage, cybersecurity systems, data processing, and custom development and test environments. Generally, platforms receive, store, change, and output the data and code by which an AI system like Affdex adds value. Typically, these capabilities run on servers located in server farms which companies like Affectiva either own or lease from providers. One can think of servers as being composed of interconnected and interoperable units of hardware (data storage systems, networking systems, and data processing systems) that make an application viable.

To function reliably, numerous workers with different jobs belonging to different occupations are responsible for designing, organizing, and maintaining platforms. Changes in the contexts to which Affdex will be applied have direct and significant implications for how the platform is configured. This is perhaps most salient for custom development and test platforms. Development and test platforms are literally assembled and changed according to the demands of those who are creating, testing, and maintaining an application like Affdex. Researchers rarely consider the labor involved in building and managing platforms when they assess the employment implications of using Affdex in a context such as border security: They focus only on end users, in this case, border guards.

Subsystems lie at the next layer down the stack. These are the technologies that make platform technologies like a server farm operable. For a single server, subsystems would include a suite of memory chips, printed circuit boards (PCBs), processors, cooling systems, displays, input/output (I/O) ports, and power units (i.e., the physical core of a computer). When it is time to field Affdex for a border security application, computing subsystems must be purchased (or at least leased), assembled, transported, configured, and maintained. In an on-demand production economy, this work is

often directly triggered by a customer order for an Affdex implementation. At any rate, some of the subsystems involved in such an
implementation would be customized for that kind of use, work
that would most certainly not have occurred without a specific type
of demand.

Below the subsystems lie the *components* that make up each subsystem. For instance, as Figure 3.3 shows, a server farm's cooling
system requires water, ductwork, fans, pumps, and heat sinks. In
any business-scale AI application such as Affdex, other components
include chips, fiber-optic cables, solid-state drives, network
switches, cameras, and so on. The elements of each class of compo-
nent are nearly identical and are usually mass-produced. Making
components reliably requires significant capital expenditure on
industrial-grade equipment such as lithography machines, robotic
transfer equipment, and lathes.

The implications of a system like Affdex for work and employment become more evident when we consider components that
have to be designed and manufactured to application-specific
specifications. For example, five years ago Google asked what would
happen if every user of a mobile device searched for images for just
five minutes every day. It found that running Tensorflow (Google's
primary machine learning software discussed in the next section)
would require between five and ten times more data centers and
power usage than currently available (Levy, 2012). Accordingly,
Google invested aggressively to create tensor processing units
(TPUs), customized chips designed to facilitate matrix multiplica-
tion, the core mathematical operation in most machine learning.
Producing TPUs required NVIDIA and other chipmakers to make
significant changes in jobs and their organization's configuration.[11]
These, in turn, filtered down to alter work practices, jobs, and work
processes of the companies that fabricated the chips.

As in the case of railroad equipment, the bottom two layers of
Affdex's material stack involve *extracting* and *refining* raw materials

---

[11] See http://www.moorinsightsstrategy.com/why-nvidia-is-building-its-own-tpu/, accessed
January 3, 2020.

from the natural environment. Chips, PCBs, plastics, and other subsystems and components are made of refined materials such as silicon, petrochemicals, lithium, iron, steel, and nickel, all of which must be extracted from the natural environment. Computational systems also require steady electrical power obtained through networks of refinement and extraction that produce either direct (photovoltaic arrays) or indirect (hydroelectricity) sources of electrical current transmitted through power lines and stored in batteries.

The extraction and refinement processes required for digital technologies, including intelligent technologies, have vast implications for work and employment. Mining and refinement comprise at least 14 percent of US GDP[12]. Power generation and distribution represent 5 percent of GDP[13]. Taken together, extraction and refinement require a massive amount of labor that cuts across numerous complex organizations, geographies, and nation-states. To be sure, the mining, refinement, and making of the components required to enable a particular application such as Affdex represent a small percentage of industrial output and GDP. Nevertheless, one should not presume that the use of intelligent technologies has no significant implications for lower levels of the material stack. For example, Gary Dickerson, CEO of Applied Materials, recently estimated that, without fundamental changes in materials, chip manufacturing, and design, the data centers that support AI applications could grow to account for as much as 10 percent of the world's electricity use (Giles, 2019). In 2016, global data centers consumed 3 percent of the world's production of electricity which is 40 percent more than all of the electricity consumed in the United Kingdom.[14]

---

[12] https://nma.org/wp-content/uploads/2016/09/Economic-Contributions-of-Mining-in-2015-Update-final.pdf, accessed January 3, 2020.

[13] https://mjbradley.com/sites/default/files/PoweringAmerica.pdf, accessed January 3, 2020.

[14] https://www.forbes.com/sites/forbestechcouncil/2017/12/15/why-energy-is-a-big-and-rapidly-growing-problem-for-data-centers/#2988edb95a30, accessed January 3, 2020.

The Digital Stack

Figure 3.4 illustrates Affdex's digital stack. Fortunately, we can use the same conceptual structure to unpack a digital stack as we did with the material one. The digital stack of an intelligent system also consists of platforms, subsystems, components, refinement processes, and extraction processes.

The development, testing, and deployment of applications like Affdex proceeds through a digital *platform*. As connectivity has improved and as software has grown more complex, organizations have created software environments that allow multiple individuals distributed across time and space to reliably and asynchronously develop, alter, and maintain the platform on which an application like Affdex runs. Platforms are sometimes made available to users as public resources, especially when applications only run on systems sold by a given vendor (IBM's Z, Intel's DevCloud). In many other cases, such as Affdex, platforms are custom-built, proprietary, and not available to the public. Most platforms are useful for a restricted range of digital applications. For this reason, applications and their platforms are deeply interdependent and idiosyncratic.

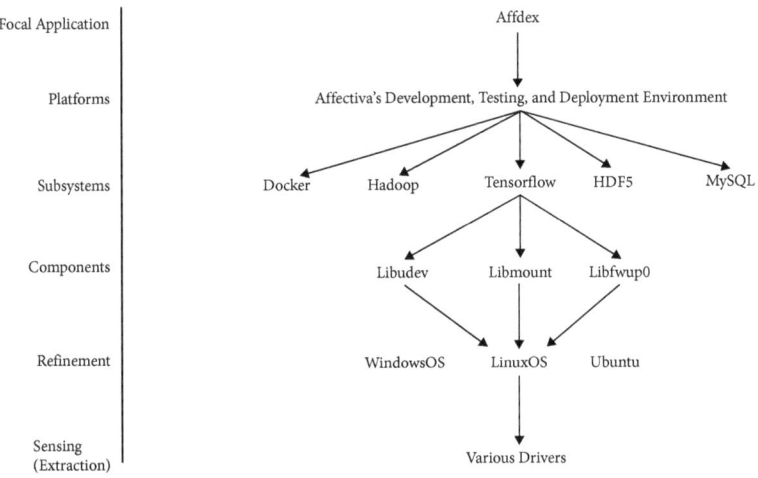

Figure 3.4. Digital Technical Base for Affdex

Digital platforms allow for contained, focused efforts to develop software and test solutions in simulated computational environments. Among other benefits, platforms reduce the probability of widespread, adverse consequences of technical errors and improve coordination between individuals and groups working on different subsystems with different test data. Building, maintaining and deploying platforms require considerable work because the design and operation of a platform is shaped by how the platform will potentially be deployed. Yet analysts who consider the implications of a system like Affdex for work and employment routinely ignore such labor. It is important to realize that developing, changing, and maintaining a platform are ongoing activities.

A platform is composed of a set of integrated digital *subsystems*. Each subsystem is a suite of data and code that executes a relatively specific function important to the focal application. Affdex's subsystems include Hadoop, a framework for storing data and running applications on clusters of commodity hardware; Tensorflow, which handles numerical computation in ways that facilitate machine learning; hdf5, which facilitates the management of extremely large and complex data sets; and MySQL, a relational database management system. Compared to platforms, subsystems are useful for a wider range of applications and are less interdependent with those applications.

As in the material stack, the implications of AI-enabled applications such as Affdex for labor and employment extend to the development of each subsystem. Tensorflow was first developed within Google as a custom framework for machine learning applications and was open-sourced in 2015. Many companies, including Affectiva, quickly turned to Tensorflow rather than develop such capabilities in house. The demand from such companies has since shaped the work required to update and maintain the Tensorflow code base. A particularly large-scale or novel deployment of a system like Affdex may well inspire a new code branch in the Tensorflow "repo" (an online code repository), but even normal deployments can occasion new and different work, which is done by the software engineers who contribute to this shared digital subsystem.

To build Affdex—or any other AI application—software engineers need either to create or to draw upon a set of existing *components* or system libraries. As in the material stack, system libraries are a collection of nonvolatile computational resources that that allow a wide range of applications to function.[15] Common examples here include libudev, which provides APIs (application programming interfaces) to enumerate and introspect (examine the internal state and functioning of) devices on the local system; libmount, which reads and manipulates filesystem tables that account for the ways that data is stored; and libfwup0, which supports managing updates to firmware (software coded into nonvolatile memory that enables hardware to function).

Libraries often need to be installed and maintained to allow a higher-level application like Affdex to operate. Components are more generic tools than applications and subsystems and are typically not deeply interdependent with focal applications. This does not mean analysts may safely ignore related labor. Because components are somewhat specialized, they require tailoring so they can interface with and support changes in an application.

Operating systems—software that supports a computer's basic functions—*refine* raw digital data so that the data can be processed up through the stack of components and subsystems in a way that an application can use. In Affdex's case, multiple operating systems are required, but the core code of components is often written for Linux, a collectively maintained, open-source operating system. Affectiva and many other applications rely on Linux because it is free, thrifty with computational resources, secure, easily custom-ized, and comes with network and server-related drivers. Other aspects of Affdex's functioning draw on the Ubuntu operating system. This is a Linux distribution that is often used for desktop computing applications and is better suited than Linux for certain drivers and hardware. Finally, given that Microsoft has expended significant effort to ensure clear standards for reliable driver

---

[15] In a digital context, nonvolatile resources are those encoded in hardware that retain their state even after the hardware running them has lost power (e.g., flash memory, read-only memory).

development, Affdex's data can partially be created through the Windows operating system.

It is difficult to account for the work required to build, maintain, and deploy operating systems that refine data. In the case of Linux, measuring labor is complicated because the software is developed through volunteer efforts over relatively long timescales. Even with Linux, however, we can get a rough sense of scale: The latest estimate of the cost of replacing the Linux kernel (core software with complete control over the entire operating system) was approximately $3 billion, which represents just over 17,000 person-years' worth of skilled work (Sz, 2011). The figure is far smaller than the iterative work that has been required to produce and maintain Linux over the years. As of 2016, Windows represented over $8 billion in revenue for Microsoft—clearly the result of a significant amount of ongoing labor.[16] Moreover, Linux and Windows have been altered to accommodate the computational processes associated with AI applications like Affdex.[17] It is important to include such labor when accounting for the work and employment occasioned by an application like Affdex.

Finally, any digital application like Affdex cannot function without converting analog data into digital signals. This *extraction* is achieved through software drivers that interpret the output of various sensors. In some cases, sensors such as keyboards, mice, and touchscreens comprise familiar human interfaces, but in other cases, sensors include cameras, microphones, and accelerometers. Most drivers are not given specific names, but people who work on them describe themselves as coding "close to the metal," which indicates that they understand that certain aspects of a digital application ultimately operate on atoms, not bits. The information extracted from sensors by drivers must be transmitted via data networks. Examples include the software behind the Internet, cell- and satellite-based telecommunications, and the cloud.

Creating digital extraction technologies requires massive interconnected technological networks. Traditional (e.g., GDP) measures

---

[16]   https://www.fool.com/investing/2017/06/29/how-microsoft-corporation-makes-most-of-its-money.aspx, accessed January 3, 2020.

[17]   https://ubuntu.com/kubeflow, accessed January 3, 2020.

of our economy do not treat digital extraction as a distinct category, so it is not easy to assess the volume and kinds of work required to enable and maintain extraction processes. Compared to the amounts of labor required further up the digital stack, employment in data extraction should constitute a smaller proportion of the work required for a business-scale AI such as Affdex. Yet, as with material extraction, we argue it is not appropriate simply to presume this work away: for example, labor involved in data transmission (e.g., Verizon workers) would likely be implicated in organizational changes associated with business-scale AI deployments.

It is important to realize that no element of a technical stack is likely to remain static. Over time technologies in the stack are likely to change, and some may leave, while others enter. Both are the outcomes of technological obsolescence. The possibility of change throughout the stack over time complicates the analysis we are proposing. For example, although a focal technology can alter the nature of work and employment of those who create the technologies that lie further down the stack, developments further down the stack can affect the work of those who create and use the focal technology. In other words, developments that shape work and employment flow up as well as down a technical stack. Recall that trains could be manufactured at scale only after several key advancements in refinement technology occurred. Affdex was coded only after improvements in processors, memory chips, algorithms, and camera design occurred. Actively considering the ways in which new technologies emerge and ramify through a technical stack thus acknowledges the reality that more people and more work may be involved in the emergence of a focal technology than initially meets the eye.

Role Systems View

Research on technology and work experienced a sea change during the 1980s as researchers moved away from more deterministic accounts of technological change toward the notion that a technology's implications for work are situated and socially constructed.

The shift led researchers into the field to do ethnographic research on how people used technologies (Barley, 1986; Orlikowski, 1992). Close, careful, long-term observation of "technologies-in-practice" (a term Orlikowski (2000) coined) not only highlighted that people and organizations could use the same technologies in very different ways, thereby affecting work and organizations in different ways, but that technologies did more than alter skills and tasks. They also sometimes altered roles, role relations, and network structures. Because roles, role relations, and networks were the subject of Chapter 2, we won't belabor the point further in this chapter except to say that because ethnographic studies focus on the micro-social dynamics of technology and work, they serve as an important corrective to reductionistic views of work and deterministic images of technological change.

Because no one has documented how Affdex is used in any context, we shall draw on field studies of other intelligent technologies to illustrate how such technologies can occasion changes in role systems. At present, only a handful of researchers have examined the situated use of intelligent technologies. Despite being few, these investigations have already begun to challenge the utility of treating work with intelligent technologies in a reductionistic way. Their message is that the changes that intelligent technologies occasion in work and employment are more varied than studies that overlook context typically presume.

Shestakofsky (2017) investigated an online labor platform that employed machine learning to match buyers and sellers of a large variety of local services such as gardening or plumbing. The company envisioned the program as saving labor costs by having an algorithm rather than a person mediate the transaction. But the program was incapable of handling the large volume and complexity of the requests it received. In response, the company hired hundreds of contractors in the Philippines to do manually what the software was supposed to do automatically. The company also developed an intelligent technology to deal with buyers and sellers who complained about the company's services. Again, the machine learning algorithm proved incapable of adequately handling the task for which is was designed. Hence, the firm decided to hire contractors in Las Vegas (mostly female) whose emotional labor proved far more effective than the algorithm at placating angry buyers and sellers. After Shestakofsky completed his research, the company moved the Las Vegas jobs to Silicon Valley, where they were performed by newly hired full-time staff.[18] Shestakofsky's study is important for three reasons. First, it shows that attempts to deploy intelligent technologies designed to make human labor unnecessary may fail and instead lead to the creation of more jobs done by humans who ironically replace or augment the machines. Second, Shestakofsky's research shows that the jobs that are created need not result in full-time employment, if the work can be out-sourced to contractors in the United States or elsewhere. Third, the jobs that are created need not involve sophisticated technical skills, an observation consistent with the now common argument that it is difficult to automate most service jobs.

Maiers (2017) investigated a data-driven algorithm designed to detect infections in babies in a neonatal intensive care unit before more common symptoms of infection appeared. Clinicians wel-comed the algorithm because of its potential to save lives. Maiers found that medical personnel negotiated the meaning of the algo-rithm's output on the basis of their situated knowledge of a patient, the opinions of other clinicians, and the presence and absence of other signs and symptoms. On the basis of this corpus of informa-tion, clinicians chose to "temper, filter, discount, or place trust in" the algorithm's predictions (p. 923). Although the designers of the technology envisioned the algorithm as a tool that would dictate care, in practice the information from the algorithm simply became another data point to consider when deciding whether and how to intervene in the case of a specific baby. Maiers' study is important because it shows us that in some instances intelligent technologies designed to simplify tasks become integrated into the routine prac-tices of members of interdependent occupations (in this case nurses, doctors, and nurse practitioners). In Markoff's (2015) terms, the technology augments rather than automates work. Moreover, in the case of the intelligent technology that Maiers

---

[18] Personal communication between Ben Shestakofsky and Matt Beane.

studied, no work was eliminated, no new jobs were created, and the roles of various practitioners remained largely unchanged.

In a two-year field study, Beane (2019) compared traditional urological surgeries with urological surgeries performed using Intuitive Surgical's da Vinci robotic surgical system at five hospitals.[19] He found that the way hospitals deployed the robots occasioned dramatic changes in role relations between surgical residents and attending physicians. In traditional open surgery, residents worked side by side with the attending physician, gradually learning to master their craft from more expert surgeons as they moved through their residency. Notably, the da Vinci robots had dual consoles from which surgeons guided the robot's instruments inside the patient's body using hand controllers and foot pedals as they viewed the surgical site and the instruments on the console's video monitor. With dual consoles, residents could, in theory, participate in the surgery along with the attending physician by taking control of the robot under the attending's supervision. However, Beane's observations and interviews amply document that most residents logged little surgical time on the robots—in part because they had few opportunities to develop the sensory motor skills necessary for controlling the instruments deftly enough to convince senior surgeons they could ensure the patient's safety. In robotic surgery communication between the attendings and the residents attempting to use the robot often became harsh and punitive when residents fumbled the controls. The attendings then abruptly resumed control of the operation, leaving the residents with little opportunity to learn the skills necessary for operating remotely. Such interactions stood in sharp contrast to the hushed collaboration typical between residents and attendings in open surgery.

Nevertheless, a small cadre of residents managed to become skilled at robotic surgery by circumventing the hospitals' norms and policies and engaging in what Beane calls "shadow learning." Residents who engaged in shadow learning spent their own time

---

[19] The Da Vinci surgical robot can be found at: https://www.davincisurgery.com/da-vinci-systems/about-da-vinci-systems, accessed April 27, 2020.

watching videos of robotic surgeries and practicing on Intuitive Surgical's simulators. They also went so far as to moonlight at other hospitals which also had da Vinci robots, but whose own surgeons were not as skilled. These hospitals and surgeons allowed the residents to perform robotic surgeries with very limited supervision. By gathering interview data from another fifteen top-tier hospitals, Beane showed that the way all hospitals deployed these robots altered roles and role relationships between residents and expert surgeons. The deployment patterns also changed the way surgical residents were traditionally trained. Finally, Beane's research strongly implies that in the long run robotic surgery threatens to create a winner-takes-all market for hyper-specialized surgeons while simultaneously increasing the risks for patients who cannot afford or do not have access to hyper-specialists.

Documenting and analyzing the situated use of intelligent technologies tempers the limitations of a reductionistic approach to studying how such technologies may influence work and employment. Rather than focus on skills, tasks, and jobs as units of analysis from which to make inferences about employment, situated studies of intelligent technologies-in-practice allow a role systems analysis of changes occasioned by an intelligent technology. Figure 3.5 illustrates how a role systems analysis would proceed. It begins by documenting whether, and if so how, using an intelligent technology shapes the tasks that a worker performs. If there is no significant change in the tasks associated with a work role, then to all intents and purposes the analysis ceases. Maiers's study illustrates this possibility: the algorithm designed to predict infections became integrated into the clinicians' existing work practices as simply another type of data for arriving at a diagnosis.

If a technology significantly changes a worker's tasks, then the analyst asks whether the worker's role has changed or whether new roles have arisen. If not, then the analysis ceases. In Shestakofsky's study, no existing roles changed (perhaps because the company was a start-up and had hoped that its algorithm would preclude hiring people to perform the tasks of matching and mollifying suppliers and clients), but new roles were clearly created to handle what the company's algorithms could not do.

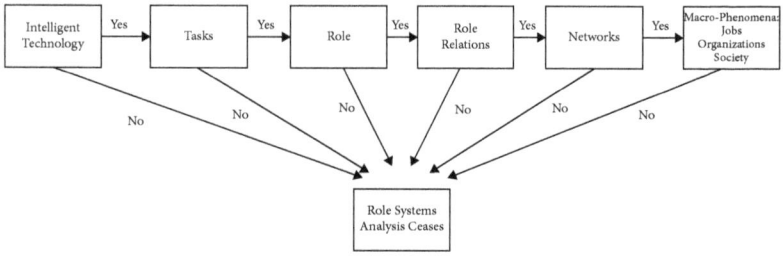

Arrows represent "associated with change in"

Figure 3.5. Role Systems Analysis of an Intelligent Technology

If roles have changed, then the analyst turns to role relations and asks whether changes in workers' roles alter their interactions with others in their role set or whether the positions or occupations comprising the role set change. As in previous steps, analysis would cease if there are there are no changes in role relations. In Beane's research, surgical robots altered both roles and role relationships in nearly identical ways across multiple field sites. Residents in robotic surgery were less likely to act as assistants than were residents in open surgery, and attendings were less likely to serve as effective mentors. The role set of some residents who engaged in shadow learning began to include physicians, nurses, and technicians in other hospitals. Hence, the networks of those residents also changed, while the networks of those who were unwilling to break norms and policies remained the same.

If networks change, analysts are at least in a position to make more grounded speculations about macro-level shifts within the occupational and organizational structures they investigate. The structure of Shestakofsky's firm clearly changed as contractors were hired to do work that the algorithm was designed to do. Shadow learning in robotic surgery created a new status structure among newly minted surgeons with a minority becoming hyper-specialized stars. Furthermore, whether patients had access to these stars depended on geography and the ability to afford the hyper-specialists' services, which reinforced the inequities of the system of medical care in the US.

Implications for Future Research on Intelligent Technologies and Work

The gist of our discussion has been that technologies, in general, and intelligent technologies, in particular, may occasion changes in work and employment that propagate both *vertically* within a technical stack and *horizontally* across the role systems of settings in which they are used. The outcomes of such propagation are also likely to vary according to a technology's specific material and digital components, the production system that creates it, and the various contexts of its use. Attempting to document and assess the implications of an intelligent technology for work and employment across such broad terrain poses significant challenges that are likely to require more eclectic methods than most students of work and employment currently use.

We have yet to find an analysis of the implications of an intelligent technology for work and employment that traces the ramifications of a focal application vertically through a technical stack. And there are relatively few that do so across multiple contexts of use. Scholars either focus on the implications of a broad class of applications covered by such terms as "robots" or "AI," thereby leveling important differences between the applications subsumed by the cover term, or they focus on the upshot of using an intelligent technology in a specific context. These approaches are respectively too broad and too narrow to document important variations and the contingences that give rise to variation. Hence, conclusions about how AI, in general, or about how a focal application like Affdex will affect jobs are likely to be incomplete, if not inaccurate. These approaches draw conclusions after investigating either the tip of the technical iceberg or the situated use of a technology in a particular setting.

Fully working out how a program of research might investigate the dynamics of work and employment in a focal application's technical stack as well as its deployment in multiple contexts is beyond the scope of this chapter. Furthermore, determining what combination of methods would be most appropriate for a task of such breadth requires the joint expertise of multiple scholars spread across a variety of disciplines from economics and engineering to the sociology of work and technology studies (Bailey and Barley, forthcoming). Nevertheless, in the remainder of this chapter we hope to open a conversation about how scholars might proceed if they were to entertain our notion of what confronts researchers who seek more inclusive and accurate answers to how intelligent technologies will change work and employment.

Investigating Technical Stacks

Although no researcher has tried to trace the implications of a focal application throughout its technical stack, supply chain analysts implicitly recognize their existence. Firms in a supply chain produce the subsystems, components, and the refined and raw materials that allow the production of a focal application. But supply chain analysis does not concern itself with issues of work and employment.

To imagine how one might begin tracing the ramifications of an intelligent technology throughout an application's material or digital stack we offer a working assumption: A focal technology is likely to have greater implications for work and employment at higher levels of a stack. The assumption has face value in the material stack because an application like Affdex is certainly less likely to have an identifiable impact on the refinement of petrochemicals for plastics or silicon for chips than on the design of the circuit boards for the servers on which Affdex runs. But the assumption also seems reasonable for a digital stack. Affdex is more likely to require a modification of Tensorflow than a modification of libmount or Linux. The assumption should limit how far down a stack an analyst would need to investigate to identify most of the major changes that a focal application might trigger in work and employment along its supply chain.[20]

---

[20] Nevertheless, analysts need to be careful when exercising the assumption that there will be more change at higher levels of a stack. For example, someone investigating the ramifications of electric vehicles for work and employment would be remiss if they did not consider

However, the assumption does not relieve researchers of the need to understand the focal technology in more depth than is usually the case in most current studies. One cannot treat a technology as a black box and still hope to trace a focal technology's implications for work and employment throughout a technical stack. This is because researchers will need to organize their data collection around platforms, subsystems, and subcomponents that have to be enumerated before they can be investigated. In other words, the nodes of a stack define the targets or contexts for study.

Identifying changes in work and employment up and down a stack should proceed more quickly if field researchers collaborate with those who use different methods to acquire data. Because few firms or informants are likely to think in terms of technical stacks, enumerate all the elements that comprise a stack, or know how a focal application might affect work and employment at all nodes of a stack, field work will be needed to uncover the structure of a stack and an application's ramifications throughout a stack. Fortunately, the necessary fieldwork need not be intensive nor the data as thick as that collected in a traditional ethnography. The objective would simply be to delineate elements of the stack and gain information on whether the focal application altered work and employment at each node. Semi-structured interviews should be sufficient for this task, as long as informants are chosen carefully to ensure that data is collected from individuals who have a broad purview of the range of technologies and production processes involved as well as from individuals who perform the work at key nodes.

Because technical stacks are generally organized as trees, one might modify Spradley's (1979) ethnosemantic techniques for mapping the structure of linguistic domains (which are also typically organized as trees) to map technical stacks. The researcher would treat the focal application as a cover term for a domain and then elicit the domain's taxonomic structure by tracing relations of inclusion such as "X is a part of Y?" or Y contains X?"[21] In context,

how the spread of such vehicles would affect the demand for lithium for manufacturing batteries.

[21] See Spradley (1979) for other question frames for eliciting the elements and structure of a tree.

researchers might map inclusion by asking questions like "What system libraries do you rely on? Who produces and maintains them?" or "What components do you use to build that tool? From whom do you acquire them?" Interviewers would first elicit the elements of what we have called platforms and then proceed to enumerating the subsystems comprising the platform. One would use the same technique to map the elements of each successively lower level of the stack. The task would most likely involve collecting data from people in organizations in an application's supply chain.

Having mapped the hierarchy of a technical stack, investigators would then turn to determining how the focal application shapes how the elements of a layer are configured or produced and whether employment has shrunk or grown as a result of these changes. One could approach this task by conducting fieldwork in organizations throughout the supply chain. But doing so is likely to be too time-consuming and labor-intensive. At the very least, researchers would have to negotiate access to multiple organizations. And, as most researchers know, gaining access is difficult enough with a single organization. For these reasons, researchers might experiment with methods that could complement or even substitute for field research.

For example, it is now possible to scrape massive volumes of job description data from the Web at high frequency. Scraping, coupled with machine learning techniques, might allow researchers to analyze apparent changes in the bundles of tasks that comprise jobs—and the work-related interdependencies between jobs. These techniques could conceivably be applied to the study of technical infrastructures via scraping job descriptions related to organizational and product infrastructure (e.g., product engineer, supply chain manager) and patent data regarding changes in technologies relevant to a material or digital stack. Analysis of this data may offer insight into the bare empirical facts associated with the vertical stack for a given technical application: of what it is comprised, how the composition changes over time, and how organizations acquire and deploy human labor to build and maintain a particular technology. Of course, such data is an imperfect representation of changes in tasks, jobs, and employment, but analyzing it could offer insight at scales unattainable via traditional qualitative techniques.

Studying Role Systems

In contrast to studying changes throughout a technical stack, students of technology and work have had more experience documenting how technologies can occasion changes in roles and role relationships in organizations that use them.[22] One might, therefore, argue that we simply need more ethnographers of work and technology to focus on the ramifications of intelligent machines. Although more studies like those conducted by Shestakofsky (2017), Maiers (2017), and Beane (2019) would be welcomed, ethnographers of work and technology are scarce. There are simply not enough fieldworkers with an interest in technology and work to execute enough studies to provide a basis for strongly inferring why intelligent technologies might occasion similar or different outcomes across different settings.

The scarcity is amplified by traditional approaches to doing ethnographies of work. Most ethnographers focus on a single context to gather rich longitudinal data on work performance as well as data on how workers interpret their working conditions and relationships. They ask such questions as who does what, with whom, under what conditions, with what technologies, and perhaps most importantly, how do people make sense of what they do and why they do it. Although such studies are extremely informative, they cannot directly address the larger issues that have made intelligent machines a topic of widespread concern. For instance, do intelligent technologies change the nature of work in a general direction, how many new occupations and jobs will be created as a result, and how much of the workforce will be left behind if intelligent technologies become common? Addressing such questions implies speaking to change in the occupational structure and the division of labor over a wide swath of the economy. Ethnographies alone cannot answer such macrostructural questions.

Nevertheless, as illustrated by the studies we reviewed in the last section, ethnographies can illuminate when and why intelligent technologies sometimes affect role systems and at other times do

---

[22]   See Leonardi and Barley (2010) for a review of such studies.

not. When designed for this purpose, ethnographies can surface the commonalities and differences that a technology may occasion in tasks, roles, and role systems across sites or contexts. But to explore commonalities as well as variation, ethnographers of technology and work need to break the tradition of studying single sites and instead embrace comparative research designs. Specifically, we need fieldwork that encompasses multiple instances of the same occupation, the same work, and the same technology. Only with multiple sites can ethnographers hope to document which changes in roles systems are common and which are not. Even more importantly, unless ethnographers study multiple sites, they cannot offer grounded accounts of why commonalities and differences occur.

Previous ethnographies of work and technology usually have emphasized variations from the expected consequences of technological change. Although understanding variation in how technologies change role systems is crucial, identifying commonalities is perhaps even more critical if ethnographers are to help identify the kinds of changes that we can expect a specific application or type of intelligent technology to occasion. The only way for a new technology to have pervasive effects on society is through common outcomes. Even though few ethnographies of technology and work stress commonalities, notable exceptions include Orlikowski and Scott (2014), Barley (1990, 2015), and Beane (2019). Orlikowski and Scott studied how a wide variety of hotels, hoteliers, and hotel workers in the UK responded by reconfiguring their internal feedback processes in response to Tripadvisor's ranking and rating system. Barley showed how selling cars through the Internet altered relationships between salespersons and customers similarly in a Toyota and a Chevrolet dealership. And as discussed in the section "Role Systems View" above, Beane documented commonalities in how residents in fifteen teaching hospitals in the US learned (or failed to learn) robotic surgery involving the DaVinci surgical system.

All of these studies examined the ramifications of the same technology among members of the same occupation in different organizations. Ethnography can also be useful for surfacing similar shifts in tasks, roles, and role relations occasioned by different technologies employed in different kinds of organizations or occupations. One way for ethnographers to surface commonalities (as well as differences) across technologies, organizations, and occupations is to design research programs to focus on several technologies, organizations, and occupations and then have a team of ethnographers divide the job of studying different technologies and contexts. This was the approach that Barley and his collaborators adopted to studying commonalities across the work of various kinds of technicians (Barley and Bechky, 1994; Barley, 1996; Barley and Orr, 1997, Barley, Bechky, and Nelson, 2016). Another approach would be to analyze the data and findings of several independently executed ethnographies. Taking this tack, Beane (2019) documented how different intelligent technologies created similar barriers to on-the-job learning in online education, platform-based local labor markets, investment banking, law enforcement, and surgery. Although it was not a study of technology per se, Hodson (2001) produced the most extensive research of this type in his analysis of the meaning of dignity at work as documented in 108 cases found in 86 ethnographies written by others (see Hodson, 2001: 299–302 for a description of his methods).

Although ethnographies are particularly well suited for studying how technologies shape role systems, we cannot rely solely on fieldworkers to amass an understanding of when and why intelligent technologies occasion changes in tasks, roles, and role relations. Even if the supply of fieldworkers were abundant, time remains an issue. Careful ethnographies require long stints of fieldwork and for this reason narrow the territory that can be covered per unit of time. Temporal constraints become more troublesome when studying change which requires collecting data before and after a technology comes on line. The temporal demands can be shortened if fieldworkers can find settings in which work is done both with and without the technology. Those who wish so study the implications of intelligent technologies for role systems would benefit from less time- and labor-intensive methods. Several possibilities that preserve the granular detail and causal insight so hard won in ethnographic work come to mind, although we have not employed them ourselves.

Pentland (Pentland, 2016; Pentland, Recker, and Kim, 2017; Pentland, Recker and Wyner, 2018) has argued that work processes can be conceptualized as a series of activities that are organized as a sequence of events that may involve either the same or different people, actions, technologies, locations, and so on. These events can be displayed as a network in which the nodes are not people (as in most network studies), but events described as an n-tuple of attributes, for example a specific person performing a specific action with a particular technology. Edges or lines in such networks represent handoffs. Pentland has demonstrated that such networks yield more complex depictions of work processes than networks constructed solely on the basis on who hands off work to whom. He argues that such network are more accurate depictions of a work process. They yield the "essential thread that holds together the fabric of organizing" (Pentland, Recker and Wyner, 2018). Pentland has designed a software package, ThreadNet, which analyzes work processes composed of work events defined by n-tuples of attributes of any length.[23]

ThreadNet may offer researchers a way to more quickly analyze how a technology occasions change in a role system in a specific context. The data which ThreadNet processes can be collected not only by field observation but from digital records or logbooks that capture who used what technology to perform what action in what sequence. When an organization automatically captures such data in databases, ThreadNet can be used to analyze changes in work processes before and after the adoption of an intelligent technology. If interpretations of insiders are not crucial to the research question, ThreadNet could augment or substitute for the thick data of an ethnography. In other words, ThreadNet may in some situations reduce the time required for an ethnography of how a technology affects a role system.

Sociometric sensors are another promising tool for studying whether and how intelligent technologies occasion changes in role

---

[23] ThreadNet can be downloaded from http://routines.broad.msu.edu/ThreadNet, accessed January 4, 2020.

systems. Sociometric sensors are devices worn or carried by individual workers that gather a wide variety of fine-grained data on workplace behavior, most notably interactions with others in the normal flow of work. The most common sociometric sensor suite is the modern smartphone. Technologies designed explicitly and solely for this purpose typically take the form of a small "badge" that can capture high-frequency data on speech volume and prosody (pitch variation) via a microphone, proximity to others via Bluetooth or infrared sensors, and bodily movement (e.g., fidgeting, respiration rate) via an accelerometer. When workers in a workplace consent to wear such badges, researchers can accumulate massive fine-grained data sets that offer significant qualitative and quantitative data on workplace dynamics. Researchers have used sociometric sensors to gain insight into expertise, social networks, work performance, and information system use (Ara et al., 2008; Fischbach et al., 2010; Waber et al., 2011). Because they can be deployed at significant scale and require little ongoing involvement from researchers (at least compared to ethnographic methods), these tools could be used to gather and perform a preliminary comparative analysis of role relations and role sets before and after an organization deploys an intelligent technology.

Text Thresher is a platform for distributed, crowd-based analysis of massive qualitative data sets such as those comprised of newspaper articles, reports and other textual data (Adams, 2015; Zhang et al., 2018). Researchers at University of California, Berkeley developed Text Thresher for a study of the "99%" protests in the United States. Attempting to understand the dynamics at play between the public, the protesters, law enforcement, the courts, and politicians, Adams realized that there was far too much data to code via traditional means (e.g., graduate student labor). Inspired by crowdsourcing platforms such as Mechanical Turk, Adams and his colleagues built an online system that allows numerous distributed individuals to examine textual (and soon image and video) data, tag the data with prescribed analytical codes, and cross-validate each other's work.[24] In an age when a wide variety of stakeholders create

---

[24] For a video describing how Text Thresher operates in more detail, see https://goodly-labs.org/tt.html, accessed April 27, 2020.

and update truly massive online data sets about the implications of intelligent technologies for their personal work (e.g., blogs, social media posts, video, court proceedings, job postings, online résumés), tools like Text Thresher may offer students of technology and work traction in studying changes in work and employment.

To envision how researchers might employ multiple methods to gain more traction on how an intelligent technology might occasion change in a role system, consider a before and after study of the use of Affdex at border security checkpoints. The research team would include a small group of ethnographers who rotate through a representative sample of checkpoints. Rotating ethnographers among checkpoints would counter differences in the fieldworkers' attention. Because the actions and decisions of immigration officers are tracked by computers, the team might also use ThreadNet to map and then compare the work process before and after Affdex arrives. If border guards, the public, and those who write about customs and border protection (CPB) create documents (discussion boards, job postings, job descriptions, articles, etc.) about using Affdex at checkpoints, the researchers might find Text Thresher useful for coding such data for insights or for checking hypotheses raised by data acquired by ethnographers. If border work is either sequentially or reciprocally interdependent, sociometric badges and smartphone data could uncover whether Affdex affects patterns of interaction among CPB personnel. Securing informed consent for collecting such data would be daunting, but not necessarily infeasible, particularly if one were to work with CPB on the research. Additionally, if CPB videotapes interactions between border guards and people crossing the border, researchers using Text Thresher might be able to code tapes to uncover changes in the tenor of interactions before and after Affdex.

Epilogue

Our agenda has been to explain why we believe we need to treat the current buzz about how intelligent technologies will change the nature of work and employment skeptically. Our intent has not been to depreciate the research that has been done. Any research is better than no research. Rather, we have tried to make three points. First, there is no reason to believe a priori that any broad class of intelligent technologies (i.e., robots, machine learning, algorithms, self-driving cars, and so on) will occasion consistent changes across settings. The accumulating body of research on other digital technologies has identified few overarching consistencies, especially across contexts of use. With respect to digitization at large, the most we can say is that the spread of computers has contributed to the elimination of many routine manual and service jobs, but even this trend has not been monolithic (Autor and Dorn, 2013).

Second, as long as researchers adopt an isolationist view of intelligent systems and a reductionist view of work, they are unlikely to uncover differences or similarities between the changes that different intelligent technologies may spawn across contexts of use. Even more importantly, they will have difficulty speaking about why those differences and similarities occur.

Third, pursuing studies of how intelligent technologies occasion changes in work and employment with an embedded view of a technology and a role systems view of work will not be easy. Nevertheless, it does not seem impossible to achieve, though doing so is likely to require a shift toward programs of research that involve scholars from a variety of disciplines as well as a variety of research methods. Fortunately, the time may be ripe for moving toward more programmatic studies of intelligent technologies and work. In this regard, the burgeoning interest in how intelligent technologies will shape our lives and our societies is a boon. The National Science Foundation in the United States has announced that understanding changes in work and employment is a key priority.[25] A number of consortia have also recently been founded to explore social and ethical questions about how intelligent technologies should be designed and deployed. These organizations include

---

[25] For the National Science Foundation's interest, see https://www.nsf.gov/news/special_reports/big_ideas/human_tech.jsp, accessed May 31, 2019, and https://www.nsf.gov/pubs/2020/nsf20515/nsf20515.htm?WT.mc_id=USNSF_25&WT.mc_ev=click, accessed December 12 2019.

Data and Society,[26] the Partnership on AI,[27] the United Nations' AI for Good,[28] the Future of Life Institute,[29] and Data Science for Social Good programs at universities.[30] Technology firms and their leaders are involved in many of these organizations, which also have significant industry funds. The success of these endeavors points to the fact that there are now funding opportunities for programmatic research that go beyond governments and foundations.

In this chapter we have ignored larger sociocultural forces that may also influence the ramifications of intelligent technologies for work and employment. In particular, we have not explored how designers' images of workplaces and work shape the constraints and affordances of a technology. Nor have we considered how the interests and ideologies of those who have the power to commission the building of an intelligent technology or who make the decision to adopt such a technology can shape the technology's ramifications. A number of scholars have shown that power, economic interests, and sociocultural dynamics are crucial for fully understanding how technologies may alter our lives, our work, and our societies (e.g. Winner, 1980; Noble, 1984; MacKenzie, 1990; Forsythe, 1993; Thomas, 1994). The fact that we have not discussed power, ideology, and culture in this chapter does not mean that we do not see them as important. Indeed, it may be that these forces are more germane to understanding intelligent technologies' implications for the future of work and employment than the material capabilities, constraints, and affordances, of the technologies themselves.[31]

Nevertheless, research on work and employment needs to also consider the material base of intelligent technologies, how the technologies are used *in situ*, and how both ramify throughout workplaces. The consequences of isolationist and reductionist views seem particularly important in this age of intelligent machines. People with power are currently relying on isolationist and

---

[26] https://datasociety.net/, accessed April 27, 2020.
[27] https://www.partnershiponai.org/, accessed April 27, 2020.
[28] https://ai4good.org/, accessed April 27, 2020.
[29] https://futureoflife.org/, accessed April 27, 2020.
[30] For the first such program, see https://dssg.uchicago.edu/, accessed April 27, 2020.
[31] For a discussion of how one might study these issues more systematically, see Bailey and Barley (forthcoming).

reductionist perspectives as they make consequential decisions regarding matters such as education, law, organization, and technological design, not to mention economic policy and jobs. Given the potential for these technologies to reconfigure interpretation, choice, and interdependent action for large and increasing segments of the population, the upshots of poor decisions are likely to be particularly consequential and troubling. For this reason alone, adopting a technologically embedded and role systems perspective may help avert the most dire scenarios that intelligent technologies could occasion.

4

Managing the Fears of Studying Technical Work

Stephen R. Barley and Diane E. Bailey

Social scientists devise research methods for many reasons. Textbooks tell us that we develop new methods to ensure the accuracy of our inferences, to make certain our observations are replicable and valid, to control heteroscedasticity, to increase the power of our inferences, to handle abnormal distributions, and so on. In this chapter we shall explore another reason, one that is rarely mentioned in methods texts or courses, but one we suspect is all too familiar to many researchers: to manage our fears. Fear is particularly prominent when researchers decide to study work that is far from their everyday experience and substantive training, as when we set out to study technical occupations and organizations. We shall show how fear, and fear alone, drove us to develop many of the methods we have used in our various studies of technical work.

Fear in research comes in many forms. Consider, for instance, a fear that haunts survey researchers. For lack of a better word, let us call it "P-fear." P-fear bears a family resemblance to how one must feel when playing Russian roulette, albeit with less dire consequences. The scenario goes like this: You have developed a hypothesis, say, "The use of communication technologies is positively associated with work-related stress and work-family conflict." You carefully develop a survey and populate it with well-vetted measures shown to have high reliability and validity. You randomly select a sample of respondents to keep other well-known research monsters at bay. You double-check your coding to make certain the data are not erroneously entered into a database. You run a factor

Work and Technological Change. Stephen R. Barley, Oxford University Press (2020). © Stephen R. Barley and Diane E. Bailey. DOI: 10.1093/oso/9780198795209.003.0004 analysis to confirm that your measures are not mixing apples with oranges and you run reliability tests on your scales to confirm that all values are above 0.7. You look at the distributions of your dependent and independent variables so that you can select just the right regression technique. Everything looks fine. You write some SAS or R code and test it to remove all bugs. Now, like the Russian roulette player spinning the cylinder, you type "run." The output quickly appears on your screen: $P \leq 0.50$. BLAM! The pin hits the shell. So much for eight months of work! So much for that paper! So much for tenure!

One of the many good things about becoming an ethnographer is that you get to say goodbye to P-fear. Choosing to become an ethnographer is initially freeing and exciting. You have chosen the final frontier. Your mission will be to explore strange new worlds, to boldly go where no researcher has gone before. No more sitting behind computer screens tossing out outliers to see if you get a better result, looking for pesky interaction effects, or splitting your sample into subsamples. Ethnography's lure is not simply freedom from P-fear and all the evasive measures that P-fear demands; there is also romance. Professors who are fieldworkers knowingly exploit this two-pronged lure of freedom and romance to entice naive graduate students, thereby ensuring a steady—although admittedly small—stream of new ethnographers.

For example, in the late 1970s and early 1980s John Van Maanen began his doctoral course at MIT on organization studies (which required all students to do a short stint of participant observation) by opening with an image of Franz Boas setting out to study Native Americans of the North American Northwest. John set the scene by inviting students to imagine Boas stepping off a boat into the wilderness, suitcase in hand, equipped only with notebooks, pencils, and several changes of underwear to document the lives of the indigenous people on Victoria Island. As John spoke, one could almost hear the waves lapping at the shore and see the sunlight filtering through the conifers as loons sang in the distance. Perhaps, Barley's career would have turned out differently if John had asked students to read Boas's diaries, in which he documented his trials, tribulations, frustrations, and fears.

Boas was an anthropologist who sought to document the languages of the Kwakiutl, the Bella Coola, and other Northwestern peoples. Born and trained in Germany, he immigrated to the US. He spoke English and he had done earlier work with the Inuit on Baffin Island. But he certainly did not speak the languages of the Northwest. What if you did not know the language or the ways of the people by whom you were suddenly surrounded and there were no English-to-whatever phrase books you could use to get by? What if you had few conceptual anchors to hang onto aside from the fact that you could pretty much bet that these people would have kinship structures, marriage rituals, death rituals, creation myths, words for plants and animals, and so on? Boas's diaries give us a sense of what such an experience might be like.

On September 21, 1886, three days after Boas landed on Victoria Island, he describes one of his first encounters with the residents of a Songis settlement:

> I also went to the Songis settlement, whose inhabitants all seem to be terrible drunkards. I wanted someone to give me words, but I had great difficulty because everyone was so drunk. Unfortunately, I had taken a map with me. There was great excitement because they believed that the railroad was to be built through their reservation. Because of the map they thought I belonged to the railroad and regarded me with undeserved suspicion. Everyone asked me whether the railroad would pass through their territory. They did not become more friendly when I wanted to write down their language. Not until I became a little rough with them and showed them money did their attitude improve.
>
> (Boas, 1969: 22–3)

Not only were Boas's informants frequently uncooperative, but he often found himself so irritated (as did Malinowski, 1967) that he could not write his field notes:

> *Nov 8 (1886).* I find that my notes are very scant these days due to the slow progress I am making. I must admit that this week has been the most unpleasant one I have spent in this region. I have to run about all day, twice a day two miles to the settlement and there is very little to be

gotten there... This afternoon I was in Somenos again and could do nothing, for the strangest reason. An old man who usually tells me stories was in best form when an equally old woman interfered. He said a man had lain dead for nine days and she said ten, whereupon he became so angry that I could not get another word out of him.

(Boas, 1969: 54–5)

Boas's diaries reveal to us the irony of ethnography's lure: P-fear is nothing compared to the self-doubt, irritation, anger, and a host of psychological insecurities that emerge from their hiding place in the researcher's closet when you commit to the field. These new fears manifest themselves in questions that torment you. Why are the people I am studying, just like Boas's Northwest peoples, being so uncooperative? Do they not like me? (Some days I do not like them!) Am I going to get the data I need? What if I look stupid to them?

When you finally leave the field for your office, new fears replace the old. Will I ever make sense out of this endless, disorganized jumble of observations and conversations? How can I weave these observations and conversations into a story that will satisfy my colleagues? What if I spend all this time in the field and discover I have nothing to say? No professors ever tell you that you will ask yourself these and many similar questions, and even if they did, it probably would not help you because you would have already succumbed to the lure.

Some people might think that sociologists, particularly sociologists of work, have it easier than anthropologists on this score. After all, sociologists of work usually do fieldwork with people in their own society who ostensibly speak the same language and mostly do not drink while on the job. In the past, when sociologists of work primarily studied manual or clerical work, it might indeed have been easier for uninitiated fieldworkers to walk onto a construction site or into a mine, a factory, or an office and more or less quickly understand what people were saying and doing. Because much industrial work was physical, it was at least observable. The same was true even for clerical work prior to the advent of computers and databases. Moreover, industrial and clerical work tended to be repetitive, so if you did not understand the work practice the first time you saw it, you probably would catch on by the time you saw it twenty times. In fact, by then you might begin to notice subtle, but important variations in how people did the task and to figure out why and when people varied how they did what they did. As a sociologist of work in the 1960s, you entered the field with an armory of trusty concepts developed by industrial sociologists since the 1930s, concepts like hierarchy, status, roles, workplace conflict, resistance to authority, working to rule, making quota, goldbricking, occupational culture, and so on. Such concepts helped sociologists of work translate their informants' emic realities into analyses that spoke to other sociologists.

But as the nature of work began to change in the late twentieth century—in part, because of the spread of computational systems and, in part, because the occupational structure began to shift away from manufacturing and clerical work toward managerial, professional, and technical work—it became more challenging for sociologists to study, and sometimes even understand, the work that people do. Compared to the mid- twentieth century, work is less physical and, hence, less easily observable. The body is less involved in the execution of work than before (Zuboff, 1988). Machines increasingly do physical and even mental labor. The entities on which people operate are becoming digitized in the sense that more and more people work on, with, or through representations rather than directly on physical objects, including paper (Bailey, Leonardi, and Barley, 2012). Co-workers are less frequently co-located, which challenges our ability to observe encounters that are distributed not only across space but across time (Hinds and Mortensen, 2005; Gilson et al., 2015).

Work on screens moves quickly. People are likely to use computer programs that do things you can never see. The practices of professional and technical workers have always rested on esoteric bodies of knowledge with which most social scientists have minimal familiarity (Abbott, 1988). As technical occupations proliferate and as technical workers come to rely on ever more sophisticated tools and software, their work becomes increasingly abstract and harder to notice, much less observe and understand. Consider, for example, the gradual shift from designing buildings and automobiles at a drafting table to doing so with computer-aided design (CAD), which automates the design and drawing of structures, and finite element analysis (FEA), which simulates physical processes such as kinetic energy moving through an object. With CAD and FEA, work that formerly involved pencils, rulers, triangles, the solving of equations, and so on is now performed by algorithms working in the background invisibly within the bowels of a computer. To make matters worse, software designers, medical professionals, researchers, technicians, and engineers of all kinds communicate with languages whose words are both phonetically and semantically difficult for outsiders to hear, much less grasp, because they are so specialized and because they refer to invisible processes. When engineers are in the middle of doing their work, they might as well be talking Kwakiutl, even though it resembles English or whatever language people use in their country.

Some thirty and twenty years ago, respectively—and for better or worse—the two of us independently decided to cast our fates to the study of technical occupations and organizations, thinking that sociologists of work needed to investigate the new worlds of work that were erupting around us. Little did we know that our decision would place us in situations more reminiscent of those faced by Boas and other early anthropologists than those faced by our immediate forebears, the ethnographers of industrial society. With every project we have undertaken, we have masochistically entered ever more alien worlds in which our background knowledge is less and less relevant. In the process, we have learned that if sociologists of work are to document the day-to-day work of members of technical occupations and organizations and then bring back stories of interest to others, they are going to confront a host of new fears. We have also learned that the only way to allay the fears is to either get out of the business or devise new methods, sometimes on the fly.

In the remainder of this chapter we shall share some of what we have learned and some of the techniques we have adopted or devised, organized by the fears we have tackled more or less in the order that we confronted them on site.[1] Although our account

---

[1] One fear that we do not discuss is the fear of whether you will gain entrée to study the technical people you wish to observe. We have not created a section addressing this fear, because we see it as a fear that all ethnographers face. We would note, however, that gaining

draws on a variety of studies that we have completed over the years, we draw most heavily on our studies of structural engineers, chip designers, and automotive engineers because these settings were the most complicated, because they caused the most fear, and because they required greater methodological inventiveness.

Fear #1: Looking Stupid

The fear of looking stupid arises early and rests on an epistemological Catch-22 that threatens a loss of face in front of your informants, whom you very much hope will like you and cooperate. The fear is particularly intense if you already have a Ph.D. and everyone knows you are a professor. The catch goes like this: "I have no idea what I'm looking for because I can't study these people's work as a competent social scientist until I know what they are doing, but I don't know anything about what they are doing." Worse yet, your informants probably *will* think you are stupid and they may say so in a language that sounds vaguely like English. Not knowing any better, you are likely to nod your head in agreement. As one of Barley's informants once put it, "I thought boys from MIT were smarter than that!"

You ignore the fear of looking stupid at your own peril and the peril of your work. For example, some years ago we had a doctoral student whom we sent into the field to observe structural engineers. When the student showed us his field notes, we found that they consisted largely of such passages as "Tim worked with John for an hour in front of a computer" or "Susan talked to Will about solving a problem." When we told the student that he needed to spend more time learning what Tim and John were doing in front of their computer and what problem Susan and Will were trying to solve and why, the student proclaimed, "I am a sociologist. I only need to

entrée can be more difficult when you are studying technical workers because their time is so highly valued by their firms and because the projects on which they work are often viewed by firms as proprietary. Accordingly, you should be prepared for a long process of negotiating access. Our experience is that you will have better luck dealing with CEOs and the heads of Research and Development than if you try to enter through Human Resources.

document interactions." After months of fruitlessly trying to convince the student that being a sociologist involved more than recording who talked to whom, his field notes remained void of substantive content. We fired him.

The key questions you have to answer to confront the fear of looking stupid are: How do you parachute into a social world prepared to study the work of people who have spent years, if not decades, being formally trained in an esoteric field about which you know absolutely nothing? How do you become familiar with a body of knowledge that is difficult to learn? We both were initially naive enough to believe that having studied anatomy and having worked in a pathology lab (Barley) would be sufficient for understanding what radiologists do and that having a Ph.D. in industrial engineering (Bailey) provided enough common ground for understanding what structural engineers do. We quickly learned that taking an undergraduate course in anatomy or knowing model-building paradigms did not help. They only offered a false sense of confidence.

Bolstered by this false sense of confidence, but realizing that we probably would need more knowledge, we tried to prepare before entering the field. We approached the problem as any good social scientist would: We "read the literature." Barley spent several months in the Harvard medical library learning the principles of how cathode ray tubes, ultrasound equipment, and CT scanners work. He scanned and read papers from radiology journals, paying close attention to the medical images contained in the papers. When we set out to study structural engineers, we also read textbooks and journals. Bailey purchased copies of Mario Salvadori's (1980) *Why Buildings Stand Up* and *Why Buildings Fall Down* (Levy and Salvadori, 1992). She thought that having read Salvadori, she had the subject covered, because what else can buildings do besides stand up or fall down?

Our preparation helped steel our nerves in the face of our fear of looking stupid, but it barely got us on the playing field. Consider Bailey's first day observing structural engineers. She knew that Salvadori (1980: 93) had noted that Newton's first and third laws, in conjunction with the principle of elasticity in materials, were "sufficient to solve almost all structural engineering problems." Salvadori had explained that Newton's first law concerned translational equilibrium, which would be achieved if the sum of the forces (*f*) acting on a building equaled zero. Similarly, Newton's third law concerned rotational equilibrium, which would be achieved if the sum of the moments (*m*, or forces multiplied by their distance from the pivot point) acting on a building equaled zero. Engineers design a building's beams and columns, Salvadori had explained, to achieve translational and rotational equilibrium. Armed with this knowledge and delighted to have to know but two equations ($\Sigma f=0$, $\Sigma m=0$) to understand how engineers solved design problems, Bailey sat down to watch an engineer "size" (i.e., determine the dimensions and weight of) steel beams. After half an hour, during a lull, Bailey remarked, "I'm still waiting for you to write the sum of the forces equals zero and the sum of the moments equals zero." Puzzled, the engineer asked for clarification. "You know," said Bailey, "Newton's first and third laws." The engineer smiled and said, "I haven't thought about them since my statics class as a sophomore. Here, let me explain to you about safety factors and you'll see what I am doing."

As this example demonstrates, and as we should have guessed, we never saw anyone in the course of everyday activities doing what the "experts said" characterized technical work. But at least our preparation gave us the confidence to attempt to study technical work. After spending just one day in the field, however, we realized we had a more fundamental problem than understanding what technical workers did, a problem that manifested itself as a second fear: "I just spent the whole day in the field and I have no idea what those people said, much less did."

Fear #2: What Did They Just Say?

Like Boas, we found it nearly impossible to capture the words that radiologists and engineers were using when they talked to each other. Unlike Boas, we could not blame it on the amount of alcohol that our informants had consumed. Sometimes our problem was relatively simple: Our informants used words we heard, but could not spell or identify. For example, structural engineers talked about "fillet" welds. In the field, taking notes as the engineers talked, we contemplated various spellings for this weld, including "fill-it," "phillet," and "filet." The spelling problem seems small, but when multiplied by all the foreign words we heard, it was substantial and slowed down our note-taking as we internally debated pronunciation: "Wouldn't 'filet weld' be pronounced 'phi-lay'? But I swear he said 'fill-it.'" Worse yet was when the words were so foreign we could not manage a first stab at spelling them. Consider this early example of recorded dialogue from our field notes of two structural engineers examining the drawing of a building design, in which the words we missed seem critical to understanding their conversation:

ENGINEER 1: The plate [*he points to an area on the drawing*] ... This is a question you can ask Gerald: How do they turn here? [**missing words**]

ENGINEER 2: The concrete ... ?

ENGINEER 1: That's normal; that's thinner. [**missing words**]

ENGINEER 2: Do we have to show how the edges of the concrete have to be?

ENGINEER 1: No.

ENGINEER 2: Do we have to tell them the maximum dimension here?

ENGINEER 1: Yes.

ENGINEER 2: I will ask Gerald about [**missing words**].

Boas's descendants, the linguistic or cognitive anthropologists, would know precisely what to do if they were facing this problem. They would document words as closely as possible and then engage in a thorough analysis of linguistic domains employing both structural and componential questions (see Spradley (1979) for how this analysis can be done). The problem is that these techniques, which Barley (1983) used in his ethnography of funeral work, take an incredible amount of time and require interviewing techniques that are difficult to use. Ethnosemantic techniques are especially irritating for your informants when they are engaged in a complicated line of action, such as trying to keep one automotive part in an FEA model from penetrating another or debugging a long run of micro-processor code. It is also unwise to interrupt a radiologist in the middle of doing a special procedure or an engineer in the middle of solving a problem to ask, "What did you just say?" If you do, you quickly discover that you are making a nuisance of yourself and may soon find yourself without informants to observe.

The language used by technical professionals is composed of con-cepts more abstract than the names for colors (Conklin, 1955) or diseases (Frake, 1961) studied by cognitive anthropologists. Engineers and technical people often talk about phenomena with no obvious material referents. A noise and vibration engineer, for example, would be hard pressed to point to an automotive part and say, "That is a case of a vibration at 159.2 Hertz with 1 G of ampli-tude." He might point to a spot on an oscilloscope or play you a sound file, but these visual or aural representations would only be traces of vibration, not the vibration itself. Besides, the job of an ethnographer of work is to document what people do, why people do it, and how they make sense of what they do, not to elicit and organize vocabularies like a linguistic anthropologist. The problem was that, before we could do the job of an ethnographer of technical work, we had to understand what people were saying and what their words meant.

To address this problem, which was particularly acute with the engineers we studied, we eventually broke with the ethnographic tradition of the solitary fieldworker and began to do observations in pairs. One member of the pair documented the actions, activities, and interactions in which an engineer engaged during a stint of work. The second used a digital audio recorder to capture talk and kept notes on the terms that the engineer used to describe what he or she was doing or that occurred in conversations with other engineers. With the second set of notes we began to build glossaries in which we recorded terms. To assist in spelling and defining the terms, we bought technical dictionaries and searched technical sites on the Web. During down times such as lunch and coffee breaks, we asked engineers for definitions of terms they had previously used that we did not understand. After several weeks of working in pairs, we were usually able to capture the technical terms that engineers most frequently uttered and our field notes began to resemble complete sentences and records of turns at talk. Compare this excerpt from later in our notes to the earlier excerpt above. Here, two engineers are examining a building diagram on a computer screen:

ENGINEER 1: The reality is the actual effective prestress in your outer spans here [*he points on the screen*] will be higher because you haven't accumulated the losses at that point. As you go towards the center, you keep accumulating losses, so the center spans are going to be a lower effective prestress, and get higher and higher.

ENGINEER 2: Well it's higher and you may have some add strands and the end-spans, anyway, because you need it to be higher because you're coming up to the center of gravity of that slab there, for one thing, so you've got a lower eccentricity and then it's wingspans.

ENGINEER 1: Right. Yeah. You follow that? [*he asks the junior engineer with them*] You know what we mean by that?

Although our notes were now complete, these methods had simply turned us into more or less accomplished parrots: We could hear and mimic technical talk, but we had no idea what the talk meant. The junior engineer in this example may have followed, but we did not. Comprehension, or more particularly semantics, was now our main fear.

Fear #3: No lo comprendo

Why did our informants utter the words they did, and what were they doing that led them to utter those words? To develop an understanding of the work, we had to be more than good mimics; we had to truly understand what our informants were saying. Thus, the fear of semantics reared its head: No lo comprendo! To deal with the fact what engineers meant passed over our heads even though the words now registered with our ears, we needed methods for learning about the practical knowledge our informants possessed as well as their tasks and tools. This point is crucial: When studying technical work, you must understand tasks and tools to get at social action because social action in technical settings revolves around doing tasks and using tools.

When studying workstation technicians and systems administrators, Barley was particularly lucky because maintaining an IT infrastructure was crisis-driven and, hence, episodic. Moreover, workstation "techs" would sometimes bring computers back to their desks to diagnose and fix, and "sys admins" usually solved problems with networks and email systems from their own desks using their own computers. During these periods of time, away from the watchful and anxious eyes of the users who were experiencing the problems, the techs and sys admins could take the time to explain terminology, what they were doing and why. Similarly, when not actually in the middle of doing a study, radiological technologists and radiologists had time to tutor Barley on how they operated technologies and how to interpret films and scans, respectively.

Finding time for informants to tutor was more difficult when studying engineers because the engineers worked for firms in which projects had deadlines and time was money. Moreover, managers insisted that we not disrupt the engineers or slow productivity. But there is a better reason to be wary of interrupting an engineer. When solving problems, and particularly when writing code, engineers come to inhabit a microworld, what software developers sometimes call "the bubble." Avid players of video games also know the experience of being intellectually and emotionally absorbed in a microworld. When engineers "lose the bubble," it may take hours for them to get back in. An ethnographer's ill-timed question can yank the engineer from the microworld that he or she has been inhabiting to solve a problem. If you break your informant's bubble enough times, your informant will be madder at you than their manager.[2] Consequently, when studying engineers, we had to develop techniques and methods to learn what engineering terms

---

[2] Kidder (1981) offers the best description of the experience of being in an engineer's microworld that we have read.

meant, how technologies worked, and why engineers did what they did without unduly interrupting their work. We sought to quell our "no lo comprendo" fear in several ways.

To begin, we purchased or borrowed from our informants training materials, technical books, and software and hardware manuals that we had seen in their cubicles. These materials spoke directly about the tools and techniques the engineers were using, and we studied the materials when not in the field. We discovered that the automobile company we studied had a library of training videos and slide decks. Accordingly, we watched the videos and studied PowerPoint presentations that explained particular technologies and techniques that our informants used in their daily work. When studying structural engineers and chip designers, we sought out engineers who were either retired or who were not members of the firms we were studying. We persuaded these engineers to serve as tutors who walked us through the steps of various tasks that we observed engineers do in the field. On at least one occasion, we found it extremely useful to invite our tutor to a bar, ply him with alcohol, ask questions, and audio-record his accounts of how he did certain tasks and how his tools worked. As we began to learn what the work entailed and what was going on when our informants used specific tools, particularly software programs, we developed an entirely new fear about recording the complexity of technical work in our field notes.

Fear #4: How Will We Capture the Complexity?

To give you a sense of how complexity manifests itself in technical work, consider the chip designers and automotive engineers we studied. Each chip designer tended to use two computer screens placed side by side on his desk. Furthermore, the designers typically divided each screen into four quadrants. Each quadrant represented a software application or a program that was running in the background and that might change with updated results. Unlike observing a physical task, the engineer and the ethnographer must monitor as many as eight lines of action simultaneously. Although each automotive engineer we studied usually worked with a single screen (because their IT policy was stingier), these engineers also usually had multiple windows open on their screen, and they flipped back and forth between the windows rapidly. As if this were not enough, email notifications and text messages popped up from time to time, as other engineers attempted to communicate with the engineer being observed. For both chip designers and automotive engineers, each window represented a different task, and behind the window lurked the complexity of specific data sets and algorithms that left only traces on the screen written in symbols, graphs, and cryptic abbreviations that glossed the information they represented. How does a researcher capture, much less understand, multiple simultaneous streams of representations that whiz on and off screens? Worse yet, informants could launch a new program in a window with as little as a keystroke.

Solving this problem was one of the most difficult empirical challenges we have ever confronted. Our techniques for doing so evolved over time in trial and error fashion. The result was an obsessive-compulsive approach to developing field notes that focused simultaneously on capturing the engineers' actions and interactions as well as how their tools worked in sufficient detail to understand what the engineers were doing with them.

We built on an approach to note-taking that Barley first used in his study of radiologists. We shadowed a single informant on multiple days for a period of four hours. We varied our observations by time of day to ensure that we captured differences in what the informant did at different times of the day. Our objective when taking notes was to capture the chronological flow of the behavior, talk, and interactions that we observed. We took notes in spiral notebooks and we wrote on the front and back of each page.[3] We sought to become a "human camera," which is, of course, impossible. Even the best fieldworker cannot see or hear everything. Action flows far more quickly than your hand can write and talk more quickly than your ears can hear. Moreover, as an observer you

---

[3] Spiral notebooks are important because the spirals allow quick flipping and remove bulk at the top of the pad that can impede note-taking or, worse yet, cause you to drop the pad.

are always oriented to a particular place in time and space, which implies that you have only one physical perspective or angle on the unfolding action. Even if you could function as a video camera, you would miss something that was obscured or outside your field of vision. Nevertheless, operating under the metaphor of being a camera was useful because it dictated that we document as much as possible. Working in this way ultimately yielded sets of field notes that recorded actions and conversations in the chronological order in which they occurred.

We augmented field notes daily with other sources of data. When taking notes, we developed the discipline of recording the time of day in ten-minute intervals. These timestamps eventually enabled us to estimate the duration of events, actions, and interactions. Because engineers and other technical workers often spend so much of their day working at computer screens, we asked our informants to take screenshots from time to time, when their computers displayed something especially important, interesting, or confusing to them or to us. Informants saved these screenshots as files, which they emailed to us at the end of the observation. We carried small digital audio recorders, which we activated when informants explained what they were doing, when they engaged in technical conversations with others, or when they attended formal meetings. When possible, we asked for copies of Excel, PowerPoint, or Word documents that the informant worked with or developed while we observed. In the case of Excel files, we asked informants about the data already contained in the file, the data they entered, the formulas that were used to process the data, and what was being calculated. When engineers used physical objects at work, we made representations of the objects by sketching them in our notebooks. We rarely took photographs because in the firms we studied managers were wary of cameras. Their fear was reasonable. We had no intention of unwittingly photographing designs that might give competitors access to a trade secret. We found that if we needed documents, photographs, or videos that illustrated the technologies or techniques that our informants used, there were usually images of previous technologies or from previous design activities that we could use which the firms did not any longer consider proprietary.

To ensure that we understood what we had seen over the course of an observation, we initially asked *our informants* to recap for us at the end of an observation what had they had been doing. We quickly found that they cast their summaries in the same mysterious terms that they had used during the course of the observation, which we had not understood the first time we heard them. Thus, we switched to debriefing the observation by recapping for the informant what *we* thought had happened. This switch turned out to be quite useful because, in listening to us, the informants saw that our understanding was incomplete and often erroneous. Accordingly, they realized they needed to correct our understanding by explaining what had gone on in less technical terms and at greater length to fill in technical background, the nature of their tasks, and, most importantly, why they had taken a course of action in the first place. With time, as our knowledge grew, we reverted to having the engineer recap the observation, and later we needed no recap at all.

In earlier studies we had learned that it was crucial to expand our day's field notes by completing sentences and writing in a narrative structure before going to sleep at night. The narrative structure was critical because it made comprehending a day's observation less difficult to understand weeks or months later. Moreover, a narrative written in complete sentences made field notes easier for collaborators to read and comprehend. Although Barley had been able to expand his notes during his research on radiologists and computer technicians in the evening after an observation, when studying engineers we found that it was impossible to observe and then complete a full set of notes in a single day. This was especially true if we wanted to include transcriptions of conversations, screenshots, and other artifacts into the flow of the notes.

Therefore, after leaving the field, we began expanding our field notes in the evening as had been our earlier practice. The expanded notes then became the scaffolding on which we elaborated by inserting transcriptions of the audio recordings at the point where the talk had occurred. We then created an appendix to each day's field notes documenting every artifact we had collected and assigned it metadata signifying the day on which it was collected and the order in which it appeared in the narrative. We then annotated each day's field notes with a description of each artifact in its appendix, including what the artifact was and who had produced it. For example, we explained graphs, tables, numbers, and abbreviations displayed in the output of a computational analyses. We noted on screenshots, drawings, and other artifacts where informants had pointed when conferring with a colleague or presenting in a meeting. Having created the appendices, we returned to our expanded notes and entered the name of each appendix at its proper place in the narrative, often elaborating the narrative with relevant information that was contained in the appendix. With all of this effort, we discovered that it took us two and half days to write up a four-hour stint of observation.

Because technologies were so important to the tasks that engineers performed and because the technologies were often specialized and unfamiliar to us, we inventoried all of the technologies that our informants used. We organized these inventories as tables in Word files, regardless of the day on which the technologies were observed. The tables contained the name of the technology and the tasks for which the technology was used, and we updated the tables as we learned more. As we compiled the inventory, we spoke with informants about how the technology had been acquired, what technologies others used to do the same task, why informants often used multiple technologies in the process of doing a task, how much leeway they had in choosing the tools they used, and how choosing one tool influenced the other technologies they would use (as, for example, when one tool's output was the input for a subsequent tool). We entered this information into the inventories as well. Compiling technology inventories enabled us to better understand how engineers understood and organized their work as well as to track the division of labor between engineers because work processes were often divided by who used what tools.

After completing the field notes for a day's observation, we reread the notes closely before we shadowed the informant again. We asked ourselves such questions as what happened, why did it happen, who was involved, and why were they involved? We read the notes to refresh our memories of what we had seen and, more importantly, to pinpoint passages in our notes about actions, tools, and interactions that we did not feel we adequately understood. Upon returning to the field, we took issues that needed resolving to our informants for further details and clarifications. Doing such "member checking" (Creswell and Miller, 2000) of the raw data in our field notes was particularly important because recording and understanding the details of work are crucial for an accurate accounting of work.

Like every ethnographer, we knew we were unlikely to make use of such an extensive corpus of notes, but we also knew that the level and accuracy of the documentation were critical for translating details into sociologically interesting and valuable insights. Moreover, unless we worked toward a detailed understanding of the technical work we were studying, at some point we would not have been able to go forward. Yet the time required to create our expanded field notes, multiplied by the number of observations we conducted, engendered our next fear.

Fear #5: How Will I Finish the Fieldwork Before I Die?

Our studies of technical work involved many, many days of observation, because, believing in the importance of comparison as the source of empirical and theoretical inspiration, most of our research has entailed studying more than one organization or occupation. Barley (1986, 1990) collected data on various imaging modalities in two hospitals when studying radiological work. His team's research on technicians was based on extensive periods of participant observation with members of seven different occupations (Barley, 1996). Bailey's fieldwork entailed studying structural engineers in three firms and chip designers in three firms. While our research on automotive engineering occurred within a single firm, we did ethnographies in three separate engineering groups: noise and vibration, body, and crashworthiness.[4] Comparative studies mean, by definition, multiple occupations and organizations to observe and, in practice, much fieldwork to accomplish.

Accomplishing that fieldwork in comparative studies raises the question of whether to study the settings simultaneously or sequentially. Because both choices arguably require the same total time, in both cases you have the fear of never completing the research, or at least of not completing it in time to publish enough papers and books for tenure or promotion. Had either of us attempted to complete the three engineering studies on our own, it is possible that death would have come more quickly than completion, which is particularly true in Barley's case because he is older than Bailey. If you are going to conduct comparative ethnographies, you have little choice but to jettison the romantic notion of the sole ethnographer embedded deeply in his or her research site. You are going to need a team.

Although ethnographers have occasionally collaborated to study the same site (Becker et al., 1961; Strauss et al., 1964), few have attempted to collaborate across settings. For this reason, there are no established guidelines for how ethnographers should pool data, much less coordinate its acquisition. In lieu of exemplars, we had to invent and refine approaches of our own. Making the choice to use a team of ethnographers raises new fears, not the least of which is how do you assure yourself that everyone will collect similar data and how will you ensure that you will understand the data after members of the team go their separate ways? Demanding that all team members collect and record data according to the tactics outlined in the previous section manages the problem of later comprehension, but the issue of comparability remains. Achieving comparability depends, in part, on how you deploy the team.

---

[4] Noise and vibration was a functional group tasked with understanding why a vehicle made noises or vibrated audibly when on the road and then eliminating the source of the problem. Body was a group tasked with creating plans for how parts would fit together while meeting specifications and suggesting modifications to designs when the parts did not fit. Crashworthiness was a group that simulated automobile crashes to determine whether the vehicle would respond adequately in traffic accidents. They confirmed their simulation results with data gathered by physically crashing a prototype vehicle into actual obstacles.

Specifically, will members of the team study different settings or will they study the same setting simultaneously?

When studying technicians, Barley took the first approach.[5] Multiple studies were undertaken at the same time and each involved six to twelve months of participant observation. Each member of the team took responsibility for one occupation and worked as a traditional ethnographer. A key feature of this organization was its approach to what ethnographers call emic versus etic analyses. Emic analysis portrays a way of life from the perspective of participants. The telling usually centers on concepts drawn from the native's worldview. In contrast, etic analyses draw more heavily on an analyst's perspectives, using social scientific theory to present a portrayal that is interesting and fruitful. Some ethnographers argue that one cannot pursue emic and etic objectives simultaneously because the two require different approaches. What is not widely appreciated is that this trade-off may be an artifact of the tradition of solitary ethnography.

Figure 4.1 displays the research design that Barley's team adopted to pursue both emic and etic objectives by separating responsibility for the two. Members of the team collectively took responsibility for comparative, etic analyses, thereby freeing individual ethnographers to pursue an emic understanding of an occupation. Barley reasoned that it was preferable for each ethnographer to work emically to ensure that any similarities identified across occupations would less likely be artifacts of the team's agenda.[6] Team meetings became the key to coordination in both data collection and analysis.

---

[5] The following description of Barley's approach to team ethnography draws heavily on Barley (1996).

[6] The occupations included sonographers employed by two urban hospitals; programmers, workstation technicians, and network administrators employed by a large university; science technicians working in monoclonal and plant pathology laboratories; medical technicians employed in a hospital laboratory; engineering technicians; and emergency medical technicians employed by two commercial and one volunteer agencies. The occupations were selected as exemplars of scientific, medical, and technical support work. For a more detailed description of the sites and methods for each study, see Barley (1986, 1990) on sonographers, Barley and Bechky (1994) on laboratory technicians, Scarselletta (1992, 1997) on medical technicians, Nelsen and Barley (1997) on emergency medical technicians, and Zabusky (1997) and Zabusky and Barley (1996) on computer technicians.

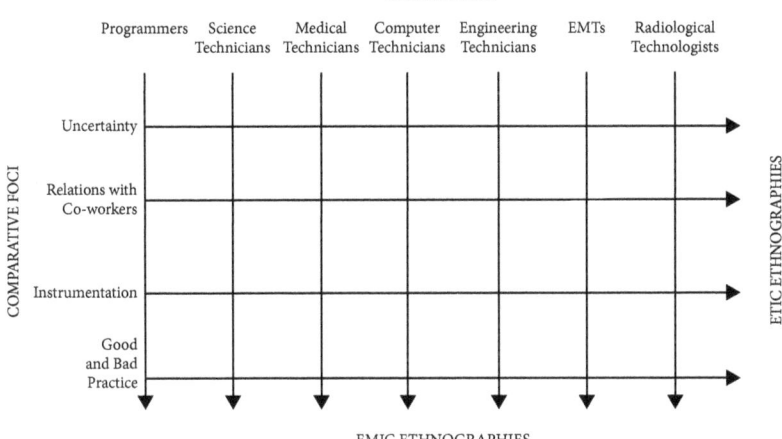

Figure 4.1. Barley's Matrix Design for Studying Technicians' Occupations

After S.R. Barley (1996) "Technicians in the Workplace: Ethnographic Evidence for Bringing Work into Organization Studies," *Administrative Science Quarterly*, 41: 404–41.

While in the field, the researchers participated in weekly team meetings, which served as fora for surfacing potential similarities and differences in the technicians' social worlds. Early meetings focused on providing the team with descriptive knowledge of the occupations under study. The team devoted each meeting to discussing one or more occupations on the basis of its collective reading of the field notes that individual researchers had compiled. The ethnographer responsible for the occupation presented a detailed portrait of the occupation's work as he or she currently understood it and shared with other members of the team insights or puzzling incidents that had occurred in the field. Team members, in turn, raised empirical and interpretive questions cued by the notes and the researchers' presentations. The objective was for the team to develop familiarity with the details of practice in various settings while beginning to identify similarities and differences among the occupations.

After gaining familiarity with each other's sites and notes, the team shifted from imparting descriptive knowledge to examining topics pertinent to all occupations under study. As illustrated in Figure 4.1, representative topics included uncertainty, relations with co-workers, instrumentation, and the technician's definition of good and bad practice. At this point, comparative analysis began. Once the team identified a potentially fruitful topic, members culled from their own notes all passages that seemed relevant to the topic and then annotated the passages to provide sufficient information for others to understand the context in which the events occurred. Each ethnographer duplicated and distributed the annotated passages. Team members read the annotated passages and developed potential coding schemes. Once everyone had processed the passages, the team met as a whole to discuss them.

The objective now became joint analysis and the production of analytic memos that detailed commonalities and differences across occupations with respect to the topic in hand. The investigation of a topic spanned multiple meetings and involved iterative coding and analysis. After joint discussions, team members reread and recoded passages prior to the next meeting, using codes on which the team had agreed. The objective was to develop an ever more refined framework for rendering the data. For instance, when examining the role of uncertainty in technicians' work, the team initially focused on identifying types of uncertainty, the coding of which led to the realization that types of uncertainty varied across settings. The team then turned from identifying types of uncertainty to identifying sources of uncertainty, consequences of uncertainty, and strategies for managing uncertainty. The end product was a series of analytic documents that detailed a mini-theory of uncertainty in technicians' work.

When studying structural engineers, chip designers, and automotive engineers, we took the second approach: All members of the research team studied the same occupation at the same site at the same time. Our teams ranged from three to four observers at any point in time. In structural engineering and chip design, Bailey served as a prominent team member, conducting fully one-half the observations at each of the six combined firms, with graduate students conducting the remainder. Her immersion in data collection across all six firms and both occupations ensured our understanding of the work. But in automotive engineering, such commitment to fieldwork on her part was not possible. Thus, we expanded our team beyond graduate students to include undergraduate students as well as R&D staff at the automotive company.

To ensure that all members of the team (including those from the automobile company) understood the project, what was expected of them, and the protocols they were to follow in producing their field notes, we developed a two-week training program. During the training program we taught basic interviewing and observational skills and had team members read and analyze field notes from previous research projects. We also had each team member observe informants who held jobs on campus, such as administrative assistants and IT technicians. After each observation, team members produced a set of expanded field notes following the procedures outlined in the previous section. The two of us read each set of field notes and annotated them with questions and comments, flagged points where information was lacking, and indicated where screenshots and other artifacts would have been helpful. We then sent the observers back into the field to fill in their earlier field notes and to conduct additional observations, which we also read and critiqued. In addition, we required that team members read and comment on each other's notes.

We had learned from our observations of structural engineers and chip designers that it was easier for team members to orient to other's field notes if the notes began with summary information on the observation. Thus, we required that all field notes begin with a header that indicated the date on which the notes were taken, the name of the person being shadowed, the name of the observer, the hours of the day during which the observation occurred, and a list of the technologies that the informant used over the course of the observation, including pencils, papers, machines, and software. Following the header, the observers wrote a narrative summary (usually a paragraph or two) that described what the informant did over the course of the observation and any important events that transpired.

Even after training the members of our team, we found it difficult to accept the fact that we were not the person collecting the data on automotive engineers. To assure ourselves that members of the team were collecting the best data possible, we adopted the practice of reviewing each observer's elaborated notes. As during training, we annotated the notes with questions designed to help the observer learn to focus on details, remember to capture additional data, and inquire about what informants were doing and why. For example, we wrote such comments as "What was on the screen at this point in time?," "What does this word mean?," "Why did she do this?," and "Who else was in the meeting?"

A team of ethnographers working simultaneously at the same site can quickly produce a large quantity of notes along with accompanying appendices and artifacts. It never occurred to us when we wrote the grant proposal that funded the project to budget in a project manager, an archivist, or a database manager. Without people in these roles, it fell to us to collect, organize, and maintain all the data; review and comment on field notes; track multiple versions of field notes in the course of our reviews; ensure conformity to standards; and monitor progress in the field. In total, our research team consisted of twenty-seven people who conducted across more than a decade observations of and interviews with engineers in three occupations, nine engineering groups, seven firms, and eight countries. We logged more than 1,000 hours of observation of more than 100 engineers and wrote thousands of pages of field notes documenting what we observed. In addition, we brought more than a thousand artifacts back from the field for further study. We maintained one complete set of hard-copy notes that filled thirty-three two-inch binders; this set included the originals of all documents and other artifacts that we took from the field. From our digital files, we created databases of notes for use in Atlas.ti^{TM}, the qualitative analysis software application that we employed in our analyses. We also created an Access database for scans of all documents and other artifacts. As you might guess, the quantity of our data prompted our next fear.

Fear #6: How Will I Make Sense of All These Data?

Perhaps the main reason we felt inundated by the quantity of data was that we had to wait until we had accumulated experience with a second occupation before we could begin to make sense of the work of the first occupation. Ideally, as methods texts suggest (Lofland and Lofland, 1984), it is smart to begin analysis before you complete your data collection. Barley was able to do so because he collected data on two radiology departments and several technicians' occupations simultaneously; in that study, coding and analysis were comparative from the beginning. But because we collected data on engineering occupations sequentially, we struggled to make sense of our data from the first occupation (structural engineering) until we felt versed in the second (chip design). For example, only after we had spent several months among chip designers did we begin to realize that structural engineering was more hierarchically organized than chip design and that this difference had significant implications for the roles that various engineers played and how learning occurred in the two settings (Bailey and Barley, 2011). In short, it was only through comparative analysis that we were able to make sociological sense of the data by looking for similarities and differences across occupations and then within occupations across sites.

In most of our studies we have found episodes to be a useful unit of analysis for chunking field notes and coming to terms with the quantity of data. By episodes we mean lines of action and/or interaction that hang together as discrete analytic and behavioral units. An episode might be an instance of when one engineer taught another engineer something. Other episodes might be instances of testing a simulation model, fixing a computer problem, collecting data on a source of noise made by an automobile while on a dynamometer, and so on. In some work settings, episodes were marked by clear beginnings and ends, as in the case of a radiological examination or the replacement, reformatting, and repopulation of a hard drive. Sometimes, episodes spanned days or weeks, as might have occurred when engineers attempted to fit several parts of an automobile together in a computer model.

Identifying episodes within the bounds of our notes has long been the strategy by which we attempt to organize data. We read though the field notes, coding the text for all instances of an episode of a particular type: a case of learning, an instance of when the output of one technology became the input for another, an instance of having to reconstruct an entire hard drive, an instance of a radiologist telling a technologist what he or she sees in a film, and so on. Once we have flagged the set of episodes of a specific kind, we then begin to code the attributes of episodes, for example, who was the teacher, who was the learner, what was learned, and by what method.

Ethnographers have long debated whether one ought to count instances of episodes, codes, and other phenomena in one's field notes. We have embraced the practice of counting, in part, because counting is a way of managing our fear in dealing with such a large quantity of data. We do not see counting as inimical to ethnography, whether emic or etic in its orientation. Of course, the numbers produced by such counting do not satisfy the demands of statistical procedures and, hence, such procedures should only be applied in special instances. Counts are useful for giving the analyst and the reader a sense of the relative frequency of certain aspects of behavior and interaction. For example, because we could count episodes of teaching and learning, because we could identify who taught and learned in each episode, and because we knew the duration and relative frequency of each episode, we were able to do comparative network analyses of teaching and learning encounters in structural engineering and chip design (Bailey and Barley, 2011). The networks allowed us to confirm our sense that teaching and learning were hierarchically organized by tenure in structural engineering but unrelated to tenure in chip design. The quantitative data served to show the reader that what we said occurred, did occur, and it enabled us to draw on other aspects of our data to explain why the difference existed.

Conclusion

Over the course of this chapter we have sought to make, and perhaps have belabored, two key points. First, the decision to study technical and professional work poses challenges that prompt fears for ethnographers that are either irrelevant or less problematic for ethnographers who study manual and clerical occupations. The challenges arise from the fact that technical work is complex. It often requires one to understand or learn bodies of knowledge that are unfamiliar to most social scientists. As a result, ethnographers who undertake such studies are likely to encounter a series of fears that they need to confront by developing approaches and methods that assist them in managing, if not quite vanquishing, their trepidations. Like all ethnographers, we have remained outsiders and relative novices, if not dilettantes, in the technical work we studied. Of course, the point of ethnography is not to become an insider but to get to the point where you can be, in Agar's (1980) words, a competent "professional stranger."

One might argue that you could avoid many of the problems and fears we encountered by making use of fieldworkers trained in the technical field you are studying. If we set aside the possibility that such fieldworkers would have too much insider's knowledge to recognize what is sociologically important and interesting, our experience is that being trained in the discipline is of limited help. Several members of the team who studied automotive engineers were trained as mechanical engineers and another was a "gearhead" who had spent considerable time in an automotive shop building and repairing cars. Although these fieldworkers had useful practical and theoretical knowledge that team members trained only in the social sciences did not have, the work they studied was far different than what they had learned in school or experienced in a shop. At best, the mechanical engineers' advantage was that some terms used by automotive engineers were not foreign to them, even though the vast majority of what they witnessed was initially just as mysterious to them as it was to the social scientists.

The second point that we have attempted to underscore is that if you are going to do fieldwork in technical settings, you need to be stubborn enough to confront your fears. The trick to managing fears is to devise methods that reduce them to conquerable proportions. In fact, some fear is probably useful, because it ensures that you remain alert to what you are studying and that you do not become complacent about what you think you know. Only by constantly questioning, do you sharpen your observations and improve the data you collect. Part of remaining on your toes is the willingness and flexibility to redesign the research and invent new methods opportunistically on the fly.

The careful reader may have already concluded that our approach places primary emphasis on the study of behavior and secondary emphasis on informants' interpretations. We take this stance because we believe that actions speak louder than words, that what people do is more important than what they say. We have, therefore, long weighted the etic over the emic, even though we do not ignore or discount the emic. Since the 1970s, it has become increasingly uncommon for ethnographers to focus primarily on behavior, whether the behavior be action or talk. Consequently, those who might consider adopting our approach should be forewarned: You are going to have to work hard in your methods sections to convince reviewers and readers that what you have to say is believable and appropriate. All too many reviewers find it difficult to accept that an ethnography need not convey an entirely interpretive account or that researchers could be so obsessive as to collect such a large corpus of behavioral data. For example, when attempting to publish our paper on teaching-learning ecologies, we found that reviewers had a hard time believing that we had data detailed enough to reflect teaching and learning and that within that data we could isolate instances of teaching and learning. It was not until we drew on the work of Roger Barker (1963) to justify and frame our approach that we began to make headway.[7] In short, although we are convinced that our methods for studying technical work have merit, and we have certainly had some success in applying them, our approach is, as Barley's father used to say, "a hard row to hoe." We would add that it has been a fun and worthwhile row to hoe. We enumerated, explained, and provided strategies for managing our fears in this chapter in the hopes that others might hoe similar rows.

---

[7] Barker was a social psychologist who invented what is now known as ecological or environmental psychology. The data that he and his colleagues collected in schools and other settings they studied consisted of minute-by-minute accounts of what his "subjects" said and did over the course of a day. Barker's research was definitely fieldwork, and it was influenced by anthropology, but it was decidedly not interpretive. His methodology was so difficult and time-consuming that it was rarely emulated beyond his students. Nevertheless, would-be fieldworkers would benefit from reading his publications.

References

Abbott, A. 1988. *The System of Professions: An Essay on the Division of Expert Labor*. Chicago: University of Chicago Press.

Acemoglu, D. and Autor, D. H. 2011. Skills, tasks and technologies: Implications for employment and earnings. *Handbook of Labor Economics* 4: 1043–71.

Adams, N. 2015. Scaling up content analysis: Crowd-coding text units. *SSRN Electronic Journal*, https://papers.ssrn.com/sol3/papers.cfm?abstract_id=2617222, accessed January 11, 2020.

Agar, Michael. 1980. *The Professional Stranger*. New York: Academic Press.

Ara, K., Kanehira, N., Olguin, D. O., Waber, B. N., Kim, T., Mohan, A., Gloor, P., Laubacher, R., Oster, D., and Pentland, A. S. 2008. Sensible organizations: Changing our businesses and work styles through sensor data. *Journal of Information Processing* 16: 1–12.

Autor, D. H. 2007. *Structural Demand Shifts and Potential Labor Supply Responses in the New Century*, Boston, MA: Federal Reserve Bank of Boston.

Autor, D. H. 2019. *Work of the Past, Work of the Future*. American Economic Association's Richard T. Ely Lecture, Atlanta, GA. January 4, 2019, https://economics.mit.edu/files/16724, accessed January 11, 2020.

Autor, D. H. and Dorn, D. 2013. The growth of low-skill service jobs and the polarization of the US Labor Market. *American Economic Review* 103(5): 1553–97

Autor, D. H., Katz, L. F., and Kearney, M. S. 2008. Trends in U.S. wage inequality: Revising the revisionists. *The Review of Economics and Statistics* 90(2): 300–23.

Bailey, D. E. and Barley, S. R. forthcoming. Beyond design and use: How scholars should study intelligent technologies. *Information and Organization*.

Bailey, D. E. and Barley, S. R. 2011. Teaching-learning ecologies: Mapping the environment to structure through action. *Organization Science* 22(1): 262–85.

Bailey, D. E., Leonardi, P. M., and Barley, S. R. 2012. The lure of the virtual. *Organization Science* 23(5): 1485–504.

Barker, R. G. 1963. *The Stream of Behavior*. New York: Appleton-Century-Crofts.

Barley, S. R. 1983. Codes of the dead: The semiotics of funeral work. *Urban Life* 10: 459–81.

Barley, S. R. 1984. *The Professional, the Semi-Professional and the Machine: The Social Implications of Computer Based Imaging in Radiology*. Dissertation. Massachusetts Institute of Technology, Massachusetts Institute of Technology.

Barley, S. R. 1986. Technology as an occasion for structuring: Evidence from observations of CT scanners and the social order of radiology departments. *Administrative Science Quarterly* 31: 78–108.

Barley, S. R. 1988a. On technology, time, and the social order: Technically induced change in the temporal organization of radiological work. In F. A. Dubinskas (ed.), *Making Time: Ethnographies of High Technology Organizations*: 123–69. Philadelphia, PA: Temple University Press.

Barley, S. R. 1988b. The social construction of a machine: Ritual, superstition, magical thinking and other pragmatic responses to running a CT scanner. In M. Locke and D. Gordon (eds.), *Knowledge and Practice in Medicine: Social, Cultural and Historical Approaches*: 497–540. Boston, MA: Kluwer Academic Publishers.

Barley, S. R. 1988c. Technology, power and the social organization of work: Toward a pragmatic theory of skilling and deskilling. *Research in the Sociology of Organizations* 6: 33–80.

Barley, S. R. 1990. The alignment of technology and structure through roles and networks. *Administrative Science Quarterly* 35: 61–103.

Barley, S. R. 1996. Technicians in the workplace: Ethnographic evidence for bringing work into organization studies. *Administrative Science Quarterly* 41: 404–41.

Barley, S. R. 2015. Why the Internet makes buying a car less loathsome: How technologies change role relations. *Academy of Management Discoveries* 1(1): 5–35.

Barley, S. R. and Bechky, B. A. 1994. In the backrooms of science: The work of technicians in science labs. *Work and Occupations* 21: 85–126.

Barley, S. R., Bechky, B. A., and Nelson, B. J. 2016. What do technicians mean when they talk about professionalism? An ethnography of speaking. *Research in the Sociology of Organizations* 47: 125–61.

Barley, S. R. and Kunda, G. 1992. Design and devotion: Surges of rational and normative ideologies of control in managerial discourse. *Administrative Science Quarterly* 37: 363–99.

Barley, S. R., Meyerson, D. M., and Grodal, S. 2011. Email as a source and symbol of stress. *Organization Science* 22: 262–85.

Barley, S. R. and Orr J. (eds.) 1997. *Between Craft and Science: Technical Work in the United States*. Ithaca, NY: ILR Press.

Barley, S. R. and Tolbert, P. S. 1997. Institutionalization and structuration: Studying the links between action and institution. *Organization Studies* 18: 93–117.

Barrett, M., Oborn, E., and Orlikowski, W. 2016. Creating value in online communities: The sociomaterial configuring of strategy, platform, and stakeholder engagement. *Information Systems Research* 27: 704–23.

Basalla, G. 1988. *The Evolution of Technology*. Cambridge: Cambridge University Press.

Beane, M. 2019. Shadow learning: Building robotic surgical skill when approved means fail. *Administrative Science Quarterly* 64(1): 87–123.

Becker, H. S. 1952. The career of the Chicago public schoolteacher. *American Journal of Sociology* 57: 470–77.

Becker, H. S. 1982. *Art Worlds*. Berkeley, CA: University of California Press.

Becker, H. S., Geer, B., Hughes, E. C., and Strauss, A. L. 1961. *Boys in White*. Chicago: University of Chicago Press.

Beniger, J. R. 1986. *The Control Revolution: Technological and Economic Origins of the Information Society*. Boston, MA: Harvard University Press.

Bijker, W. E. and Pinch T. 1987. The social construction of facts and artifacts. In W. E. Bijker, T. P. Hughes, and T. Pinch (eds.), *The Social Construction of Technological Systems: New Directions in the Sociology and History of Technology*: 17–50. Cambridge, MA: MIT Press.

Blauner, R. 1964. *Alienation and Freedom: The Factory Worker and His Industry.* Chicago: University of Chicago Press.

Boas, F. 1969. *The Ethnography of Franz Boas.* Chicago: University of Chicago Press.

Boczkowski, P. J. and Orlikowski W. J. 2004. Organizational discourse and new media: A practice perspective. In D. Grant, C. Hardy, C. Oswick, and L. L. Putnam (eds.), *Handbook of Organizational Discourse*: 359–77. Thousand Oaks, CA: Sage.

Brynjolfsson, E. and McAffee, A. 2011. *Race against the Machine.* Lexington, MA: Digital Frontier Press.

Brynjolfsson, E. and McAffee, A. 2014. *The Second Machine Age: Work, Progress, and Prosperity in a Time of Brilliant Technologies.* New York: W.W. Norton and Company.

Bughin, J., Lund, S., and Remes, J. 2016. Ten new work orthodoxies for the second machine age. In B. Lanvin and P. Evans (eds.), *The Global Talent Competitiveness Index 2017*: 359–77. Fontainebleau: INSEAD, Adecco, and HCLI.

Burt, R. S. 1982. *Toward a Structural Theory of Action: Network Models of Social Structure, Perception and Action.* New York: Academic Press.

Bylinsky, G. and Moore A. H. 1985. Flexible manufacturing systems. In T. Forester (ed.), *The Information Technology Revolution*: 284–94. Cambridge, MA: MIT Press.

Calo, R. and Rosenblat, A. 2017. The taking economy: Uber, information, and power. *Columbia Law Review* 117: 1623.

Chandler, A. D. 1977. *The Visible Hand: The Managerial Revolution in American Business.* Boston, MA: Harvard University Press.

Chui, M., Manyika, J., and Miremadi, M. 2015. Four fundamentals of workplace automation. *McKinsey Quarterly* 29(3): 1–9

Cialdini, R. B. 1984. *Influence: The Psychology of Persuasion.* New York: Quill.

Collins, R. 1988. Theoretical continuities in Goffman's work. In P. Drew and A. Wootton (eds.), *Erving Goffman: Exploring the Interaction Order*: 41–63. Cambridge: Polity Press.

Collins, R. 2004. *Interaction Ritual Chains.* Princeton, NJ: Princeton University Press.

Conklin, H. C. 1955. Hanunoo color categories. *Southwestern Journal of Anthropology* 11: 39–44.

Coombs, R. W. 1984. Long-term trends in automation. In P. Marstrand (ed), *New Technology and the Future of Work and Skills*: 147–62. London: Francis Pinter.

Cowan, R. S. 1985. *More Work for Mother: The Ironies of Household Technology from the Open Hearth to the Microwave.* New York: Basic Books.

Creswell, J. W. and Miller, Dana L. 2000. Determining validity in qualitative inquiry. *Theory into Practice* 39: 124–30.

Daniels, R. 2000. *Trains across the Continent: North American Railroad History.* Bloomington, IN: University of Indiana Press.

Darr, A. and Pinch, T. 2013. Performing sales: Material scripts and the social organization of obligation. *Organization Studies* 34(11): 1601–21.

Davis, G. F. 2016. *The Vanishing American Corporation: Navigating the Hazards of a New Economy.* Oakland, CA: Berrett-Koeler Publishers.

De Smet, A., Lund, S., and Schaninger, W. 2016. Organizing for the future. *McKinsey Quarterly* 1: 30–43.

Deverell, W. G. 1994. *Railroad Crossing: Californians and the Railroad* 1850–1910. Berkeley, CA: University of California Press.

Domingos, P. 2015. *The Master Algorithm: How the Quest for the Ultimate Machine Will Remake Our World*. New York: Basic Books.

Emerson, H. 1912. *The Twelves Principles of Efficiency*. New York: Engineering Magazine Company.

Emerson, H. 1914. *Efficiency as a Basis for Operation and Wages*. New York: Engineering Magazine Company.

Faraj, S., Pachidi, S., and Sayegh, K. 2018. Working and organizing in the age of the learning algorithm. Information and Organization 28(1): 62–70.

Faunce, W. A. 1965. Automation and the division of labor. *Social Problems* 13: 147–60.

Feldman, M. S. and Pentland B. T. 2003. Reconceptualizing organizational routines as a source of flexibility and change. *Administrative Science Quarterly* 48: 94–118.

Fischbach, K., Gloor, P. A., Lassenius, C., Olguin, D. O., Pentland, A. S., Putzke, J., and Schoder, D. 2010. Analyzing the flow of knowledge with sociometric badges. *Procedia: Social and Behavioral Sciences* 2(4): 6389–97.

Fischer, C. S. 1992. *America Calling: A Social History of the Telephone to 1940*. Berkeley, CA: University of California Press.

Flink, J. J. 1975. *The Car Culture*. Cambridge, MA: MIT Press.

Ford, M. 2015. *Rise of the Robots: Technology and the Threat of a Jobless Future*. New York: Basic Books.

Forsythe, Diana E. 1993. The construction of work in artificial intelligence. *Science, Technology and Human Values* 18(4): 460–79.

Frake, C. O. 1961. The diagnosis of disease among the Subanum of Mindanao. *American Anthropologist* 63: 113–32.

Freeman, C. 1984. *Long Waves in the World Economy*. London: Frances Pinter.

Frey, C. B. and Osborne, M. A. 2013. *The Future of Employment*. Oxford: Oxford Martin Programme on Technology and Employment.

Fry, L. W. 1982. Technology-structure research: Three critical issues. *Academy of Management Journal* 25: 532–52.

Furman. J., Holdren, J. P., Muñoz, C., Smith, M., and Zients, J. 2016. *Artificial Intelligence, Automation, and the Economy*. Washington, DC: Executive Office of the President.

Galperin, R. V. 2017. Mass-production of professional services and pseudo-professional identity in tax preparation work. *Academy of Management Discoveries* 3(2): 208–29.

Garfinkel, H. 1967. *Studies in Ethnomethodology*. Englewood Cliffs, NJ: Prentice Hall.

Geertz, C. 1973. *The Interpretation of Cultures*. New York: Basic Books.

Gerwin, D. 1979. The comparative analysis of structure and technology: A critical appraisal. *Academy of Management Review* 4: 41–51.

Giddens, A. and Turner J. H. 1987. *Social Theory Today*. Stanford, CA: Stanford University Press.

Giles, M. 2019. *Is AI the Next Big Climate-Change Threat? We Haven't a Clue.* Cambridge, MA: MIT Technology Review, https://www.technologyreview.com/s/614005/ai-computing-cloud-computing-microchips/, accessed January 11, 2020.

Gilson, Lucy L., Maynard, M. T., Young N. C. J., Vartiainen M., and Hakonen M. 2015. Virtual teams research 10 years, 10 themes, and 10 opportunities. *Journal of Management* 41: 1313–37.

Glaser, B. G. and Strauss. A. L. 1965. *Awareness of Dying.* Chicago: Aldine.

Glaser, B. G. and Strauss. A. L. 1971. *Status Passage.* Chicago: Aldine Publishing Company.

Goffman, E. 1959. *The Presentation of Self in Everyday Life.* Garden City, NY: Doubleday.

Goffman, E. 1961a. *Asylums.* New York: Anchor.

Goffman, E. 1961b. *Encounters.* Indianapolis, IN: Bobbs-Merrill.

Goffman, E. 1967. *Interaction Ritual: Essays on Face-to-Face Behavior.* Garden City, NY: Doubleday.

Goffman, E. 1974. *Frame Analysis.* New York: Harper and Row.

Goffman, E. 1983. The Interaction Order. *American Sociological Review* 48: 1–17.

Gould, R. V. and Fernandez, R. M. 1989. Structures of mediation: A formal approach to brokerage in transaction networks. *Sociological Methodology* 14: 89–126.

Granovetter, M. S. and McGuire, P. 1998. The making of an industry: Electricity in the United States. *The Sociological Review* 41(S1): 147–73.

Haber, S. 1964. *Efficiency and Uplift: Scientific Management in the Progressive Era 1890–1920.* Chicago: University of Chicago Press.

Hadley A. T. 1892. *Railroad Transportation: Its History and Its Laws.* New York: G. P. Putnam's Sons.

Hage, J. and Aiken, Mi. 1969. Routine technology, social structure, and organization goals. *Administrative Science Quarterly* 14: 366–76.

Heater, B. 2019. Amazon debuts a pair of new warehouse robots. *TechCrunch*, http://social.techcrunch.com/2019/06/05/amazon-debuts-a-pair-of-new-warehouse-robots/, accessed January 11, 2020.

Hinds, P. J. and Mortensen, M. 2005. Understanding conflict in geographically distributed teams: The moderating effects of shared identity, shared context, and spontaneous communication. *Organization Science* 16: 290–07.

Hodson, R. 2001. *Dignity at Work.* Cambridge: Cambridge University Press.

Hughes, E. C. 1958. *Men and Their Work.* Glencoe, IL: Free Press.

Hughes, T. P. 1983. *Networks of Power: Electrification in Western Society, 1880–1930.* Baltimore, MD: Johns Hopkins Press.

Hughes, T. P. 1994. Technological momentum. In M. R. Smith and L. Marx (eds.), *Does Technology Drive History? The Dilemma of Technological* Determinism: 101–13. Cambridge, MA: MIT Press.

Kaplan, T. 1981. *Humans Need Not Apply.* New Haven, CT: Yale University Press.

Kidder, J. 2015. *Soul of a New Machine.* Boston, MA: Little, Brown and Company.

Kondratieff, N. D. 1935. The long waves in economic life. *The Review of Economic Statistics* 17: 105–15.

Krackhardt, D. and Hanson, J. R. 1993. Informal networks: The company behind the chart. *Harvard Business Review* 71: 104–12.

Leonardi, P. M. and Bailey, D. E. 2008. Transformational technologies and the creation of new work practices: Making implicit knowledge explicit in task-based offshoring. *MIS Quarterly* 32: 411–36.

Leonardi, P. M. and Barley, S. R. 2008. Materiality and change: Challenges to building better theory about technology and organizing. *Information and Organization* 18: 159–76.

Leonardi, P. M. and Barley, S. R. 2010. What's under construction here? Social action, materiality, and power in constructivist studies of technology and organizing. *The Academy of Management Annals* 4: 1–51.

Levy, F. and Murnane, R. J. 2005. *The New Division of Labor: How Computers Are Creating the Next Job Market*. Princeton, NJ: Princeton University Press.

Levy, M. and Salvadori, M. 1992. *Why Buildings Fall Down: How Structures Fail*. New York: W. W. Norton.

Levy, S. 2012. Google throws open doors to its top-secret data center. *Wired*, 20(11), https://www.wired.com/2012/10/ff-inside-google-data-center/, accessed January 11, 2020.

Lofland, J. and Lofland, L. H. 1984. *Analyzing Social Settings: A Guide to Qualitative Observation and Analysis*. Belmont, CA: Wadsworth.

McChesney, R. W. and Nichols J. 2016. *People Get Ready: The Fight Against a Jobless Economy and a Citizenless Democracy*. New York: Nation Books.

MacKenzie, D. 1990. *Inventing Accuracy: A Historical Sociology of Nuclear Missile Guidance*. Cambridge, MA: MIT Press.

McKinsey Global Institute. 2017a. *Jobs Lost, Jobs Gained: Workforce Transitions in a Time of Automation*. New York: McKinsey and Company.

McKinsey Global Institute. 2017b. *A Future That Works: Automation, Employment and Productivity*. New York: McKinsey and Company.

McKinsey Global Institute. 2018. *Skill Shift: Automation and the Future of the Workforce*. New York: McKinsey and Company.

Maiers, C. 2017. Analytics in action: Users and predictive data in the neonatal intensive care unit. *Information, Communication and Society* 20(6): 915–29.

Malinowski, B. 1967. *A Diary in the Strict Sense of the Term*. London: Routledge and Kegan Paul.

Markoff, J. 2015. *Machines of Loving Grace: The Quest for Common Ground between Humans and Robots*. New York: Harper Collins.

Mead, G. H. 1934. *Mind, Self, and Society*. Chicago: University of Chicago Press.

Memorial Sloan Kettering Hospital. 2012. *Memorial Sloan Kettering Cancer Center, IBM to Collaborate in Applying Watson Technology to Help Oncologists*, https://www.mskcc.org/press-releases/mskcc-ibm-collaborate-applying-watson-technology-help-oncologists, accessed January 11, 2020.

Messinger, S. L., Sampson, H. S., and Towne, R. D. 1962. Life as theater: Some notes on the dramaturgic approach to social reality. *Sociometry* 25: 98–110.

Mindell, A. 2015. *Our Robots, Ourselves*. New York: Viking.

Muro, M., Maxim, R., and Whiton, J. 2019. *Automation and Artificial Intelligence: How Machines Are Affecting People and Places*. Washington, DC: Metropolitan

Policy Program, Brookings, https://think-asia.org/handle/11540/9686, accessed January 11, 2020.

Nadel, S. F. 1957. *The Theory of Social Structure*. Glencoe, IL: Free Press

National Academies of Science, Engineering and Medicine. 2017. *Information Technology and the U.S. Workforce: Where Are We and Where Do We Go from Here?* Washington, DC: National Academies Press.

National Science and Technology Council. 2016a. *Preparing for the Future of Artificial Intelligence*. Washington, DC: National Science and Technology Council.

National Science and Technology Council. 2016b. *The National Artificial Intelligence Research and Development Strategic Plan*. Washington, DC: National Science and Technology Council.

Nelsen, B. J. and Barley, S. R. 1997. For love or money: Commodification and the construction of an occupational mandate. *Administrative Science Quarterly* 42: 619–53.

Nelson A. J. 2019. *Lessons from the international diffusion of early railroad signaling technologies*. Working Paper. Eugene, OR: Lundquist College of Business, University of Oregon.

Nelson, D. 1980. *Frederick W. Taylor and the Rise of Scientific Management*. Madison, WI: University of Wisconsin Press.

Nielsen Company, LLC. 2018. *U.S. Music 360—2018 Highlights*, https://www.nielsen.com/us/en/insights/report/2018/us-music-360-2018/, accessed January 11, 2020.

Noble, D. F. 1984. *Forces of Production: A Social History of Industrial Automation*. New York: Knopf.

O'Neil, C. 2016. *Weapons of Math Destruction: How Big Data Increases Inequality and Threatens Democracy*. New York: Crown.

Organisation for Economic Co-operation and Development. 2016. *Automation and Independent Work in a Digital Economy. Policy Brief on the Future of Work*. Paris: OECD Publishing.

Orlikowski, W. J. 1992. The duality of technology: Rethinking the concept of technology in organizations. *Organization Science* 3(3): 398–427.

Orlikowski, W. J. 2000. Using technology and constituting structures: A practice lens for studying technology in organizations. *Organization Science* 11(4): 404–28.

Orlikowski, W. J. 2007. Sociomaterial practices: Exploring technology at work. *Organization Studies* 28(3): 1435–48.

Orlikowski, W. J. and Barley, S. R. 2001. Technology and institutions: What can research on information technology and research on organizations learn from each other? *MIS Quarterly* 25(3): 145–65.

Orlikowski, W. J. and Scott, S. V. 2008. Sociomateriality: Challenging the separation of technology, work and organization. *The Academy of Management Annals* 2: 433–74.

Orlikowski, W. J. and Scott, S. V. 2014. What happens when evaluation goes online? Exploring apparatuses of valuation in the travel sector. *Organization Science* 25(3),: 868–91.

Park, R. E. and Burgess, E. W. 1921. *Introduction to the Science of Sociology*. Chicago: University of Chicago Press.

Pelto, P. 1973. *The Snowmobile Revolution: Technological and Social Change in the Artic*. Menlo Park, CA: Cummings.

Pentland, B. T. 1992. Organizing moves in software support hotlines. *Administrative Science Quarterly* 37: 527–48.

Pentland, B. T. 2016. ThreadNet: Tracing and visualizing associations between actions. Paper Presented at the IFIP Working Group 8.2, Dublin, December 2016.

Pentland, B. T., Recker, J., and Kim, I. 2017. Capturing reality in flight? Empirical tools for strong process theory. ICIS 2017 Proceedings. https://aisel.aisnet.org/icis2017/ResearchMethods/Presentations/4, accessed April 28, 2020.

Pentland, B. T., Recker, J., and Wyner, G. 2018. Rediscovering handoffs. *Academy of Management Discoveries* 3(3): 284–01.

Pinch, T. 2010. The invisible technologies of Goffman's sociology from the merry-go-round to the Internet. *Technology and Culture* 51(2): 409–24.

Powell, W. W. 1990. Neither market nor hierarchy: Network forms of organization. *Research in Organizational Behavior*, 12: 295–35.

Rostow, W. W. 1980. The long cycle: An integrated view. In W. W. Rostow (ed.), *Why the Poor Get Richer and the Rich Slow Down*: 3–60. Austin, TX: University of Texas.

Sacks, H., Schegloff, E. A., and Jefferson, G. 1974. A simplest systematics for the organization of turn-taking for conversations. *Language* 50(4): 696–735.

Salvadori, M. 1980. *Why Buildings Stand Up*. New York: W. W. Norton.

Scarselletta, M. 1992. Button pushers and ribbon cutters: Observations on skill and practice in a hospital laboratory and their implications for the shortage of skilled technicians. Working Paper. National Center for the Education of the Workforce. Philadelphia, PA: University of Pennsylvania.

Scarselletta, M. 1997. The infamous "Lab Error": Education, skill and quality in medical technician's work. In S. R. Barley and J. E. Orr (eds.), *Between Technology and Society: Technical Workers in Modern Settings*: 187–09. Ithaca, NY: ILR Press.

Schank, R. C. and Abelson, R. P. 1977. *Scripts, Plans, Goals, and Understanding*. Hillsdale, NJ: Lawrence Erlbaum.

Schegloff, E. A. 1987. Analyzing single episodes of interaction: An exercise in conversation analysis. *Social Psychology Quarterly* 50: 101–14.

Schultze, U. and Orlikowski, W. J. 2004. A practice perspective on technology-mediated network relations: The use of Internet-based self-serve technologies. *Information Systems Research* 15(1): 87–106.

Schumpeter, J. A. 1934. *The Theory of Economic Development*. Cambridge, MA: Harvard University Press.

Schwab, K. 2016. *The Fourth Industrial Revolution: What it Means, How to Respond*. World Economic Forum, https://www.weforum.org/agenda/2016/01/the-fourth-industrial-revolution-what-it-means-and-how-to-respond/, accessed January 11, 2020.

Shestakofsky, B. 2017. Working algorithms: Software automation and the future of work. *Work and Occupations* 44(4): 376–23.

Smith, M. L. 1994. Recourse of empire: Landscapes of progress in technological America. In M. R. Smith and L. Marx (eds.), *Does Technology Drive History? The Dilemma of Technological* Determinism: 37–52. Cambridge, MA: MIT Press.

Spradley, J. P. 1979. *The Ethnographic Interview*. New York: Holt, Rinehart and Winston.

Sproull, L. S. and Kiesler, S. 1991. *Connections: New Ways of Working in the Networked Organization*. Cambridge, MA: MIT Press.

Stover, J. F. 1961. *American Railroads*. Chicago: University of Chicago Press.

Strauss, A. L. 1959. *Mirrors and Masks*. New York: Free Press.

Strauss, A. L. 1978. *Negotiations: Varieties, Processes, Context and Social Order*. San Francisco: Jossy-Bass.

Strauss, A. L., Schatzman, L., Bucher, R., Ehrlich, D., and Sabshin, M. 1964. *Psychiatric Ideologies and Institutions*. Glencoe, IL: Free Press.

Susskind, R. and Susskind, D. 2015. *The Future of the Professions: How Technology Will Transform the Work of Human Experts*. Oxford: Oxford University Press.

Sz. 2011. Cost of Linux. http://linuxcost.blogspot.com/2011/03/cost-of-linux.html, accessed January 11, 2020.

Taylor, F. W. 1911. *The Principles of Scientific Management*. New York: Norton.

Thomas, R. J. 1994. *What Machines Can't Do: Politics and Technology in the Industrial Enterprise*. Berkeley, CA: University of California Press.

Thompson, J. D. 1967. *Organizations in Action*. New York: McGraw Hill.

Turner, R. E. and Edgley, C. 1976. Death as theater: A dramaturgical analysis of the American funeral. *Sociology and Social Research* 60: 377–92.

U.S. Bureau of the Census. 1975. *Historical statistics of the United States, colonial times to 1970*. Washington, DC: U.S. Department of Commerce, Bureau of the Census.

Vallas, S. P. 2001. Symbolic boundaries and the new division of labor: Engineers, workers and the restructuring of factory life. *Research in Social Stratification and Mobility* 18: 3–37.

van Duijn, J. J. 1983. Fluctuations in innovations over time. In C. Freeman (ed.) *Long Waves in the World Economy*: 19–30, London: Frances Pinter.

Waber, B. N., Aral, S., Olguin, D., Wu, L., Brynjolfsson, E., and Pentland, A. 2011. Sociometric badges: A new tool for IS research. doi.org/10.2139/ssrn.1789103.

Weber, M. 1922/1968. *Economy and Society*. Berkeley, CA: University of California Press.

Weick, K. E. 1993. The collapse of sensemaking in organizations: The Mann Gulch disaster. *Administrative Science Quarterly* 38: 628–52.

White, H. C., Boorman S. A., and Breiger. R. L. 1976. Social structure from multiple networks I: Blockmodels of roles and positions. *American Journal of Sociology* 81: 730–80.

Wikipedia. 2019. Deep blue versus Garry Kasparov, https://en.wikipedia.org/wiki/Deep_Blue_versus_Garry_Kasparov, accessed May 10, 2020.

Wikipedia. 2020. Watson (computer), https://en.wikipedia.org/wiki/Watson_(computer), accessed May 10, 2020.

Williams, R. C. 2008. The maintainers of safety and efficiency: The brotherhood of railroad signalmen, 1900–1940. Master's Thesis, Department of History, University of Maryland.

Wilshire, B. 1982. The dramaturgical model of behavior: Its strengths and weaknesses. *Symbolic Interaction* 5: 287–98.

Winner, L. 1980. Do artifacts have politics? Daedalus 109(1): 121–36.

Wolfson, S. 2018, April 24. "We've got more money swirling around": How streaming saved the music industry, https://www.theguardian.com/music/2018/apr/24/weve-got-more-money-swirling-around-how-streaming-saved-the-music-industry, accessed January 11, 2020.

Woodward, J. 1958. *Management and Technology*. London: HMSO.

World Economic Forum. 2016. *The Future of Jobs: Employment, Skills and Workforce Strategy for the Fourth Industrial Revolution*. Geneva: World Economic Forum.

Yates, J. 1993. *Control through Communication: The Rise of System in American Management*. Baltimore, MD: Johns Hopkins.

Zabusky, S. E. 1997. Computers, clients, and expertise: Negotiating technical identities in a non-technical world. In S. R. Barley and J. E. Orr (eds.), *Between Technology and Society: Technical Workers in Modern Settings*: 129–53. Ithaca, NY: ILR Press.

Zabusky, S. E. and Barley, S. R. 1996. Redefining success: Ethnographic observations on the careers of technicians. In Paul Osterman (ed.) *Broken Ladders*: 185–214. Cambridge: Cambridge University Press.

Zetka, J. R. 2003. *Surgeons and the Scope*. Ithaca, NY: Cornell University Press.

Zhang, A. X., Ranganathan, A., Metz, S. E., Appling, S., Sehat, C. M., Gilmore, N., Adams, N. B., Vincent, E., Lee, J., and Robbins, M. 2018. A structured response to misinformation: Defining and annotating credibility indicators in news articles. *World Wide Web Conference 2018*, 603–12, https://homes.cs.washington.edu/~axz/papers/webconf_credco.pdf, accessed January 11, 2020.

Zuboff, S. 1988. *In the Age of the Smart Machine*. New York: Basic Books.

Index

Note: Figures are indicated by an italic "*f*", following the page number.

For the benefit of digital users, indexed terms that span two pages (e.g., 52–53) may, on occasion, appear on only one of those pages.

Adams, N. 111–12
administrative assistants 30, 67–8
Affectiva's Affdex 87–9, 98, 103–4, 112
   digital stack 93–7, 93*f*
   material stack 89–93, 89*f*
Agar, M. 142–3
AI *see* artificial intelligence
AI for Good 113–14
Amazon
   automation 20
   employees 22–3
   robots 72
American Federation of Labor 85–6
American Pragmatism 28
anthropologists 118–21
   cognitive 125–6
   linguistic 125–6
Apple Maps 89 n.10
application programming interfaces
   (APIs) 95
applications (intelligent technologies) 92–6
   see also Affectiva's Affdex
Applied Materials 92
artifacts in research 131–3, 138–40
artificial intelligence (AI)
   current thinking 71–2, 78–9
   future research 103
   technologically embedded view 86–96,
     89*f*, 93*f*
   see also Affectiva's Affdex
assembly lines 13–14, 18
Atlas.ti™ 140
audiences for encounters 34–5
Autobytel 57–8
automation
   current thinking 73–4, 78
   railroads 84–6
   and service jobs 98–9
automobiles *see* cars

Bailey, D. E. 65–6
Baltimore and Ohio (B&O) Railroad 80–1
Barker, R. G. 28 n.2, 144
Barley, S. R. 108–9
Beane, M. 100–2, 107–9
Beniger, J. R. 12–13, 15
big data 20–1
blast furnace 81
BMW 20 n.9
Boas, F. 117–18, 121, 124–6
border security 87–91, 112
Brandeis, L. 14
Brotherhood of Signalmen 84–5
Brynjolfsson, E. 71–3
Bylinsky, G. 18–19

cars
   design and testing 65–6
   electric 104 n.20
   floor sales 52–7, 54*f*, 61–2, 64–5
   Internet sales 57–62, 59*f*, 64–5, 67, 108
   isolationist view 75–6
   self-driving 21, 72
chess 72
Chevrolet 108
Chicago School of Sociology 28
chip designers 65
Cialdini, R. B. 55 n.26
cognitive anthropologists 125–6
coke (fuel) 81
collaborative research 126–7, 135–40, 137*f*
comparative research 134–41, 137*f*
complexity in research, capturing 129–34
components (digital stack) 93*f*, 95
components (material stack) 89*f*, 129–30
components (technical stack) 81–2, 82*f*
   railroads 82–3, 83*f*
comprehension, lack of in research
   127–9

computer-aided design (CAD) 74
  car design and testing 65–6
  control revolution 18–20
  fieldwork challenges 120–1
computer-aided manufacturing 18–20
computerized medical imaging 36–52,
    64–5, 67
  roles and role relations 49–52
  scripts 40–9
computerized numerical control
    (CNC) 18–19
computerized tomography (CT) 37, 39, 64
  roles and role relationships 50–2, 51f
computer simulation technology 65–6
confidence in research 123–4
constructionism 25–6
contingency theories 63
control revolution 1–2, 15, 18–23
  Faunce's theory 17f
  see also Third Industrial Revolution
control technologies 18–23
  Faunce's theory 15–16, 17f
  railroads 86 n.6
conversion technologies 17–18
  Faunce's theory 16, 17f
Costco 57–8
counts in research 142

data, making sense of research 140–2
Data and Society 113–14
Data Science for Social Good university
    programs 113–14
da Vinci robots 100–1, 108
debriefing in research 132
Defense of the Ancients 86–7
determinism 13, 25–6
  contingency theories 63
  intelligent technologies 97–8
Dickerson, G. 92
digital stacks 88–9, 93–7, 93f, 103–4, 106
digital subtraction angiography (DSA)
    37, 50
digitization 74–5, 112–13
  fieldwork challenges 120
dramaturgical analysis of technological
    change 32–6, 33f
driverless cars 21, 72

Eastern Railroad 14
ecological psychology 144 n.1
economic long waves 11–12
efficiency 14–15

Efficiency Society 15
electric grid
  related technologies 12–13
  Second Industrial Revolution 18
  smart grid 20–1
electric motor 18
electric vehicles 104 n.20
Emerson, Harrington 15
emic analysis 136, 137f, 142, 144
empiricism 70–3, 75, 79
encounters 28, 31–6, 62–3, 65
  car sales 53–62
  computerized imaging 37–8, 49–50
  dramaturgical elements 32–6, 33f
Engelbart, Douglas 72–3
engineers, locomotive 85–6
entrée gaining in research 121 n.1
environmental psychology 144 n.1
episodes in research 141–2
ethnography 121, 142–4
  lures 117, 119
  P-fear, freedom from 117, 119
  role systems view of intelligent
    technologies 97–8, 107–12
  see also fears in research
ethnosemantics 105–6, 125–6
etic analysis 136, 137f, 142, 144
event networks 110
extraction (digital stack) 93f, 96–7
extraction (material stack) 89f, 91–2
extraction (technical stack) 81–2, 82f
  railroads 82–3, 83f

Fairchild Semiconductor 15–16
Fanuc 19
Faunce, W. A. 15–16, 17f, 19–22,
    80, 86 n.6
fears in research 116–22, 142–3
  complexity, capturing 129–34
  comprehension, lack of 127–9
  gaining entrée 121 n.1
  language problems 124–7
  looking stupid 122–4
  making sense of data 140–2
  P-fear 116–17, 119
  time issues 134–40
Fernandez, R. M. 66–7
field notes
  Boas's 118–19
  chunking using episodes 141–2
  collaborative research 138–40
  complexity problems 129–34

counts, use of 142
ineffective 122–3
language problems 124–7
fieldwork *see* fears in research
finite element analysis (FEA) 120–1
First Industrial Revolution 1, 8 n.4, 17, 21–3
Faunce's theory 17*f*
first-order effects 10–11
fluoroscopy 36–7, 39, 44 n.16
focal application (digital stack) 93*f*, 94
focal application (material stack) 89–90, 89*f*
focal application (technical stack) 81–2, 82*f*, 97
intelligent technologies 103–6
railroads 83–4, 83*f*
footing in encounters 34–6
Ford, H. 13–14, 18
Ford, M. 72–3
Fourth Industrial Revolution 1–2
challenges 23–4
inanimate humans 21
misconception 74–5
myths about technological revolutions 11–13
frames 31–2
Frey, C. B. 73–4
Future of Life Institute 113–14

gaining entrée in research 121 n.1
Garfinkel, H. 32–3
gatekeepers 65–7
Geertz, C. 28 n.2
Goffman, E. 31–2, 34–5, 37, 61
Google
big data 20–1
Maps 89 n.10
Tensorflow 91, 94, 104
Gould, R. V. 66–7

Hadoop 94
handicraft era 16–17, 21–2
Faunce's theory 17*f*
hdf5 94
Hodson, R. 108–9
Horvitz, E. 78 n.3
Hughes, T. P. 25

IBM
Deep Blue 72
Watson 72
Z platform 93
idealism 13 n.6

inanimate humans 21, 78
infrastructural technologies 6–12, 9*f*
foreshadowing of 15
history of revolutions 16
materialistic determinism, fallacy of 13–14
railroads 80
research on intelligent technologies 106
Second Industrial Revolution 7–10, 12
second-order effects 69–70
swarming of innovations 12–13
technological revolutions 74–5
innovations, swarming of 11–13, 21, 74–5
insurance quoting systems 66–7
Intel 15–16
DevCloud platform 93
intelligent technologies 69–71, 112–15
current thinking 71–80
future research 107–12
role systems view 77, 86–7, 97–103, 102*f*, 107–15
technologically embedded view 77, 86–97, 103, 113–15
internal combustion engine 18
International Brotherhood of Electric Workers 85–6
Internet
and AI 78–9
car sales 57–62, 59*f*, 64–5, 67, 108
control revolution 19–20
Interstate Commerce Commission (ICC) 14
Intuitive Surgical's da Vinci robots 100–1, 108
isolationist view of technology 75–8
intelligent technologies 88–9, 113–15
railroads 80–1

Jeopardy 72

Kasparov, G. 72
Kidder, J. 128 n.2
Kiesler, S. 10
Kiva 72

language problems in research 124–7
law 23
Leonardi, Paul 65–6
Levy, F. 23
libfwup0 95
libmount 95, 104

libudev 95
linguistic anthropologists 125–6
Linux 95–6, 104
locomotive engineers 85–6
looking stupid, fear of 122–4
Lyft 72

McAfee, A. 71–3
McCarthy, J. 72–3
machine learning
    current thinking 72–3, 78–9
    research on intelligent technologies 106
    role systems view 98–9
    see also Affectiva's Affdex; Google:
        Tensorflow
McKinsey Global Institute 74
magnetic resonance imaging (MRI)
    37 n.11
Maiers, C. 99–101, 107
Markoff, J. 72–3, 99–100
materialism 13 n.6, 25–6
materialistic determinism, fallacy of 13–15
materials handling technologies see transfer
    technologies
material stacks 88–92, 89f, 103–4, 106
Mechanical Turk 111–12
medicine 23
    lung cancer diagnosis 72
    neonatal infection detection 99–101
    orthopedic surgery 67–8
    radiology see radiology
    robotic surgery 23, 100–2, 108
Memorial Sloan Kettering Cancer
    Center 72
microchip designers 65
microprocessors 18–19
Microsoft
    inanimate humans 21, 78
    Windows 95–6
microworlds 128–9
Mindell, A. 72–3
mobile computing 78–9
Moore, A. H. 18–19
moves in encounters 34–6
    car sales 56–7
Murnane, R. J. 23
music industry 69–70
MySQL 94

Nadel, S. F. 28–9
National Science Foundation 113–14

networks see event networks; social
    networks
Newcomen, T. 17 n.8
non-relational roles 28–30, 36, 65–6
    computerized imaging 37–8
nuclear magnetic resonance (NMR) 37 n.11
numerical control (NC) 18–19
NVIDIA 91

OECD 74
OpenAI 86–7
operating systems 95–6
organizational change 25–68
    car sales 57–62
    computerized imaging 49–52
organizational networks see social networks
Orlikowski, W. J. 25–6, 66–7, 97–8, 108
orthopedic surgery 67–8
Osborne, M. A. 73–4
outsourcing 98–9
    car design and testing 65–6
    inanimate control systems, emergence
        of 22–3
    jobs safe from 23

Partnership on AI 113–14
Pelto, P., *The Snowmobile Revolution*
    2–3, 5
Pentland, B. T. 64, 110
P-fear 116–17, 119
photography in research 131
platforms (digital stack) 93–4, 93f
platforms (material stack) 89f, 90
platforms (technical stack) 81–2, 82f
    intelligent technologies 105–6
    railroads 82–3, 83f
positions 28
positron emission tomography
    (PET) 37 n.11
power technologies 17
    Faunce's theory 16, 17f
process technologies 16
    see also conversion technologies
professors 30
props for encounters 34–6, 65
    car sales 53, 61–2
prototypes 78

radiography 36–9, 64
    roles and role relationships 49–52, 51f
    scripts 40–4, 41f, 49–50

radiology 23, 36–52, 64–5, 67
  roles and role relations 49–52
  scripts 40–9
railroads 79
  efficiency 14–15
  role systems 84–6
  technological embeddedness 80–4, 83*f*
recapping in research 132
reductionist view of work 76–8
  intelligent technologies 97–8, 101, 113–15
  railroads 84
refinement (digital stack) 93*f*, 95–6
refinement (material stack) 89*f*, 91–2
refinement (technical stack) 81–2, 82*f*
  railroads 82–3, 83*f*
reindeer herding 3–10, 13–14
relational roles 28–30, 65–6
  computerized imaging 37–8
research
  confidence 123–4
  fears *see* fears in research
  intelligent technologies 107–12
robotics
  control revolution 18–19
  current thinking 71–4, 78–9
  future research 103
  reductionist view of work 76
  surgery 100–2, 108
role relationships 26–8, 30–1, 35–6, 62–3,
    65–8
  computerized imaging 37–8, 38*f*,
    49–52, 51*f*
  intelligent technologies 75–7, 97–8,
    100–2, 107–11
  railroads 85–6
roles 26–31, 34–6, 62–7
  computerized imaging 37, 49–52
  intelligent technologies 75–7, 86–7,
    97–8, 100–2, 107–10
  railroads 84–5
role systems view
  of intelligent technologies 77, 86–7,
    97–103, 102*f*, 107–15
  of railroads 84–6

Salvadori, M. 123–4
Saturn 64
Schultze, U. 66–7
Schumpeter J. A. 11–12, 21
Schwab, K. 1
scientific management 14–15

Scott, S. V. 108
scraping of data 106
screenshots in research 131–3, 138–9
scripts 32–6, 64–5
  car sales 53–62, 54*f*, 59*f*, 64
  computerized imaging 37–8, 40–50,
    41*f*, 45*f*
Second Industrial Revolution 1, 18, 21–3, 69
  efficiency 14–15
  Faunce's theory 17*f*
  infrastructural technologies 7–10, 12
  innovations 12
second-order effects 11, 69–70
  computerized imaging 50–2
secretaries *see* administrative assistants
self-driving vehicles 21, 23, 72
semiconductors 18–19
sensors
  current thinking 72
  sociometric 110–12
servomechanisms 18–19
Shestakofsky, B. 98–9, 101–2, 107
signalmen and maintainers 84–6
Silicon Valley 1, 13
Skolt Lapps 3–10, 13–14
smart grid 20–1
smartphone data 110–12
snowmobiles 5–10, 13–14
social constructionism 25–6
social networks 26–7, 35–6, 62–3, 67–8
  computerized imaging 50–2, 51*f*
  intelligent technologies 75–7, 87,
    97–8, 102
  sociometric sensors, insights
    from 110–11
Society to Promote the Science of
    Management 15
sociologists of work 119–21
  see also fears in research
socio-materiality 11
sociometric sensors 110–12
sonograms *see* ultrasound
Spradley, J. P. 105–6
Sproull, L. S. 10
stacks *see* digital stacks; material stacks;
    technical stacks
stages for encounters 34–6, 65
  car sales 53, 56–8, 61–2
steam engines 17
structural engineers 65
stupid, fear of looking 122–4

substitutional technologies 6–7
  infrastructural technologies viewed as 10
subsystems (digital stack) 93*f*, 94
subsystems (material stack) 89*f*, 90–1
subsystems (technical stack) 81–2, 82*f*
  intelligent technologies 105–6
  railroads 82–3, 83*f*
supply chain analysis 104
supporting actors 33–4, 65
  car sales 53, 61–2
Susskind, D. 78–9
Susskind, R. 78–9

Taylor, F. W. 14–15, 77–8
Taylor Society 15
teamwork in research 126–7, 135–40, 137*f*
technical stacks 81–2, 82*f*
  digital stacks 88–9, 93–7, 93*f*,
    103–4, 106
  intelligent technologies 86–97, 103–4
  material stacks 88–92, 89*f*, 103–4, 106
  railroads 82–4, 83*f*
technological isolationist view
  see isolationist view of technology
technologically embedded view
  framework 81–2, 82*f*
  of intelligent technologies 77, 86–97,
    103, 113–15
  of railroads 80–4, 83*f*
  see also technical stacks
technological revolutions 1–24, 74–5
  history 16–23
  myths 11–15
  "revolution" term 15
Tensorflow 91, 94, 104

Text Thresher 111–12
Third Industrial Revolution 1
  see also control revolution
Thompson, J. D. 29–30
ThreadNet 110, 112
time issues in research 134–40
timestamps in field notes 131
Toyota 108
training of researchers 143
  collaborative research 138–40
transfer technologies 18–20
  Faunce's theory 16, 17*f*
  railroads 80
Tripadvisor 108

Uber 72
Ubuntu 95–6
ultrasound 37–9, 64
  scripts 44–50, 45*f*
United Nations' AI for Good 113–14

Van Maanen, J. 117
Verizon 96–7
voluntarism 13 n.6

water power 17
Watt, J. 17
Waymo 72
Waze 89 n.10
WebGA 66–7
Wilson, W. 14 n.7
Windows 95–6
World Economic Forum 1

X-rays *see* radiography